T0261344

TECHNOLOGIES FOR HOME NETWORKING

THE WILEY BICENTENNIAL—KNOWLEDGE FOR GENERATIONS

*E*ach generation has its unique needs and aspirations. When Charles Wiley first opened his small printing shop in lower Manhattan in 1807, it was a generation of boundless potential searching for an identity. And we were there, helping to define a new American literary tradition. Over half a century later, in the midst of the Second Industrial Revolution, it was a generation focused on building the future. Once again, we were there, supplying the critical scientific, technical, and engineering knowledge that helped frame the world. Throughout the 20th Century, and into the new millennium, nations began to reach out beyond their own borders and a new international community was born. Wiley was there, expanding its operations around the world to enable a global exchange of ideas, opinions, and know-how.

For 200 years, Wiley has been an integral part of each generation's journey, enabling the flow of information and understanding necessary to meet their needs and fulfill their aspirations. Today, bold new technologies are changing the way we live and learn. Wiley will be there, providing you the must-have knowledge you need to imagine new worlds, new possibilities, and new opportunities.

Generations come and go, but you can always count on Wiley to provide you the knowledge you need, when and where you need it!

WILLIAM J. PESCE
PRESIDENT AND CHIEF EXECUTIVE OFFICER

PETER BOOTH WILEY
CHAIRMAN OF THE BOARD

TECHNOLOGIES FOR HOME NETWORKING

Edited By

SUDHIR DIXIT and RAMJEE PRASAD

WILEY-INTERSCIENCE
A JOHN WILEY & SONS, INC., PUBLICATION

Published by John Wiley & Sons, Inc., Hoboken, New Jersey
Published simultaneously in Canada

For general information on our other products and services please contact our Customer Care Department within the U.S. at 877-762-2974, outside the U. S. at 317-572-3993 or fax 317-572-4002.

Wiley also publishes it books in variety of electronic formats. Some content that appears in print, however, may not be available in electronic format.

Wiley Bicentennial Logo: Richard J. Pacifico

Library of Congress Cataloging-in-publication Data:

Technologies for Home Networking edited by Sudhir Dixit & Ramjee Prasad.
     p. cm.
  Includes index.
  ISBN 978-0-470-07374-2 (cloth)
1. Home computer networks. 2. Home automation. I. Dixit, Sudhir. II. Prasad, Ramjee.
  TK5105.75.N48 2007
  004.6′ 8—dc22

                                              2007023223

Printed in the United States of America

10 9 8 7 6 5 4 3 2 1

To my brothers: Sushil, Sunil, and Sunit

—Sudhir Dixit

To my wife Jyoti, our daughter Neeli, our sons Anand and Rajeev, our granddaughters Sneha and Ruchika, and our grandson Akash

—Ramjee Prasad

CONTENTS

PREFACE

आपूर्यमाणमचलप्रतिष्ठं-
समुद्रमापः प्रविशन्ति यद्वत् ।
तद्वत्कामा यं प्रविशन्ति सर्वे
स शान्तिमाप्नोति न कामकामी ।।
(२।७०)

āpūryamāṇam acala-pratiṣṭhaṁ
samudram āpaḥ praviśanti yadvat
tadvat kāmā yaṁ praviśanti sarve
sa śāntim āpnoti na kāma-kāmī

A person who is not disturbed by the incessant flow of desires—that enter like rivers into the ocean, which is ever being filled but is always still—can alone achieve peace, and not the man who strives to satisfy such desires.

—The Bhagvad-Gita (2.70)

The home networking market took off in about 2003, when consumers opted, in large numbers, for the high-speed Internet connection over DSL or cable connection. At about the same time, the prices of WLAN access points dropped to under $100, and a vast majority of households and enterprises began to deploy WLAN at their premises. Now, access points with much higher speed can be bought for under $50. Concurrent with the WLAN deployments, large-scale commercialization of Bluetooth, ZigBee, and other short-range radio technologies are under way to provide wireless interfaces to all types of devices and equipment at home and to interconnect them in a seamless fashion. The In-Stat/MDR reports that by the year 2008, the home networking market will reach over $17 billion. Clearly, this market offers huge opportunities to the manufacturers and network providers and also to consumers to enjoy ultimate flexibility and significantly enhanced experience. In the future, it is anticipated that the private networks (e.g., home networks) would become part of the global network ecosystem, participating in sharing their own content and running IP-based services, (e.g., VoIP, IPTV), possibly becoming service providers themselves. This is already happening in the so-called social networks and peer-to-peer content delivery networks that are service-layer overlays on the Internet. Nevertheless, this trend has brought up the issues of digital rights and copyright management and security and authentication.

This book is about the latest topics in home networking, such as the use cases, various networking technologies, security, service discovery, media formats and description, media distribution, security, digital rights management, and the role of sensor technologies in the home environment. Because each topic can easily expand into a book of its own and it is difficult to have in-depth knowledge in all of these domains, we chose to invite the various experts in the fields to contribute their thoughts. The book is written in a style to provide a broad overview of the home networking technologies with a special emphasis on the user as the center of all activities in the home. The book is aimed toward practicing engineers, graduate students, and researchers. It has been our objective to provide the material in one single place to enable quick learning of the fundamentals involved in an easy-to-read format.

Finally, we (the authors and editors) have tried our best to ensure that each and every chapter is as accurate as possible; however, some errors in any manuscript are inevitable. Please let us know of any errors and ideas to improve the book—such comments will be highly appreciated.

ACKNOWLEDGMENTS

We are indebted to the contributors of this book for their hard work that made this book possible. All throughout this project, they were patient and forthcoming with any revisions we requested of them.

Sudhir Dixit thanks his wife, Asha, daughter Sapna, and son Amar, for their support and understanding while he worked long hours editing this book.

We express our gratitude to the staff of John Wiley, especially Paul Petralia and Whitney Lesch, for being patient with us as we missed deadlines several times to deliver the manuscript. They provided us invaluable help during the course of the whole publication process.

We also thank the International Wireless Summit and the Wireless Personal Multimedia Conference, held in September 2005, where the chapters published in this book were first presented as the invited talks in a special session on home networking.

SUDHIR DIXIT
Mountain View, California

RAMJEE PRASAD
Aalborg, Denmark
April 2007

CONTRIBUTOR LIST

Mahbubul Alam, Cisco Systems (maalam@cisco.com)

Sudhir Dixit, Nokia Siemens Networks, Mountain View, CA 94043, USA (sudhir.dixit@nsn.com)

John Farserotu, CSEM (john.farserotu@csem.ch)

John F.M. Gerrits, CSEM (John.gerrits@csem.ch)

Claus Lindholt Hansen, Ericsson (claus.1.hansen@ericsson)

Edwin A. Heredia, Microsoft (eheredia@microsoft.com)

Linda Källström, Helsinki University of Technology (linda.kallstrom@tml.hut.fi)

Dimitris Kalofonos, Nokia (dimitris.kalofonos@nokia.com)

Tommi Mikkonen, Tampere University of Technology (tommi.mikkonen@tut.fi)

Ramjee Prasad, Aalborg University, Aalborg, Denmark (prasad@es.aau.dk)

Franklin Reynolds, Nokia (franklin.reynolds@nokia.com)

Mika Saaranen, Nokia (mika.saaranen@nokia.com, mika.saaranen@ieee.org)

Jussi Saarinen, Tampere University of Technology (jussi.p.saarinen@tut.fi)

Juha Saarnio, Nokia (juha.saarnio@nokia.com)

Saad Shakhshir, Nokia (shakhshir@gmail.com)

Zach Shelby, University of Oulu (zach.shelby@ee.oulu.fi)

Sanna Suoranta, Helsinki University of Technology (sos@tml.hut.fi)

Henry Tirri, Nokia (henri.tirri@nokia.com)

Anthony Vetro, Mitsubishi Electric Research Laboratories (avetro@merl.com)

Paul Wisner, Nokia (paul.wisner@nokia.com)

Heather Yu, Panasonic Research Laboratories (heathery@research.panasonic.com)

1

INTRODUCTION TO NETWORKED HOME

Mahbubul Alam, Sudhir Dixit, and Ramjee Prasad

Advances in communications technology to seamlessly connect all types of home devices and appliances are driving the vision to create an intelligent home ecosystem. This would enable control, access, and information sharing among all the devices and thereby a much more enhanced user experience.

The future growth of electronics at home lies in the devices being able to wirelessly communicate among themselves and with one or more universal handheld portable multiradio devices (including other intelligent control points). Such a control device would be able to control the other wireless-enabled devices in a distributed or centralized manner. All devices would in the future come with some type of a radio interface built into them. One could potentially conceive the giant intelligent "home system" as being distributed but connected in a modular fashion over a large-area wireless infrastructure. This naturally requires a vast amount of research in various aspects of networking, from privacy/security to high performance to seamless connectivity, emulating "being there" with the device or the equipment.

Although there are currently handheld devices such as PDAs, PSPs, and iPODs, imagine one universal device (by which a person is still in control) with all of these features and more. Imagine this "wonder" device that can allow one to wirelessly connect to (and control) one or more electronic devices within one's house over an ad hoc network. Every new electronic device would come armed with an "antenna" to send and receive information and with some type of sensor built into it. The gizmo could be voice activated, touch screen activated, or be designed to take whatever input the user finds preferable and allow one to stay connected while roaming around the house, such as watching the television while moving

Technologies for Home Networking. Edited by Sudhir Dixit and Ramjee Prasad
Copyright © 2008 John Wiley & Sons, Inc.

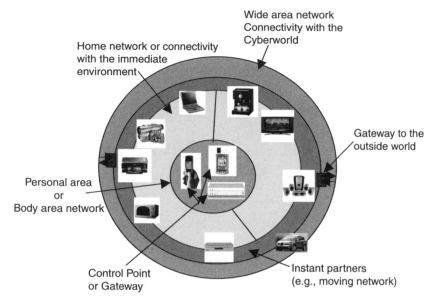

FIGURE 1.1   A general depiction of devices interconnected with a home network.

from room to room. This would also allow one to connect to one's computer and play MP3 music that is not stored in it, or download a video stream, or stream a video to a server, or share it with someone else in a peer-to-peer configuration. One could activate the coffeemaker from the bed, control the lights, watch the security video, control the thermostat, and so on and so forth. When the home network is connected to the Internet, it would also be possible to do all those things from a remote location as well either from a computer type of console or from the same handheld device as in the home.

The home network of the future would be both under the control of the human and a machine that has been trained or has acquired knowledge from the user via self-learning techniques. In short, wireless connectivity would be the key enabler to creating smart space to enhance a person's quality of life and to ease the use of the intelligent devices in his or her proximity. Figure 1.1 shows a general high-level depiction of what the future entails. Clearly, the opportunities are enormous.

In this chapter, we first provide a broad overview of consumer, technology, and marketing trends to familiarize the reader with the drivers behind developing a networked home. This is followed by a brief outline of the rest of the book.

1.1   BACKGROUND

Internet usage has approximately doubled since 2000, and in 2006, it stands at more than 1 billion people worldwide, or 18% of the world population. The rate of growth is slowing down but is expected to increase again once broadband is further developed for high-speed, rich-media content delivered at a reduced price. Major growth in the future is expected from the developing countries with large populations. Much of the growth contribution will come from wireless and collaborative applications that require access to the Internet. This has to be a wake-up call to pundits who have witnessed Internet access quickly becoming as ubiquitous as electric power, telephones, televisions, or any other public utility. It also does a lot to explain much of the current Internet hype that emphasizes non-PC applications. By applying a couple of simple rules of thumb, we can quickly ascertain where things are really headed and why. Figure 1.2 shows the mobile and broadband subscribers compared with the installed base of computers.

At a very rudimentary level, there are two ways to grow the market adoption for any major new technology:

1. By attracting early adopters followed by normal consumers.
2. Through a generational change where nonusers literally die out to be replaced by a whole new generation of consumers who are comfortable with the new technology.

Innovation can generally reach about two thirds of the market penetration through the first method, but to reach nearly 100% market share, the latter approach is required. That is the way it has always been even though people choose to forget that fact. That is why electricity and telephone, both of which have been with us for more than a

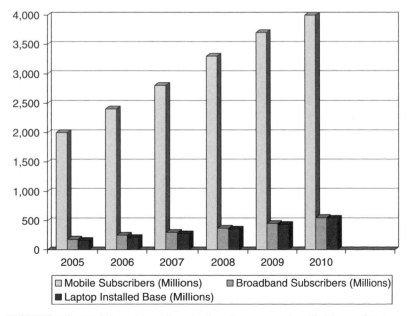

FIGURE 1.2 Mobile and broadband subscribers versus installed base of laptops.

century, have almost 100% market penetration in most developed countries. People who are born today cannot imagine being without a telephone or electricity. The only communication product class to violate this intergenerational trend is broadcast television, which grew to above 90% market penetration in less than 30 years. That is most probably because television was perceived as an extension of radio by consumers and therefore maintained a much longer effective adoption cycle.

Today, the Internet and cable television use in the United States are roughly comparable at just under 70% market penetration. That means the commercial Internet, which effectively dates from the late 1980s, has grown at about three times the rate of cable TV, which began in the late 1940s and took until 1976 to reach 15% penetration. In fact, cable TV market penetration stood at 50% in 1987, about the time the commercial Internet came into being. Therefore, the Internet has grown a lot faster than these earlier communication technologies, but then the Internet is technically dependent, for the most part, on some other host networks. At the very least, a consumer needs to be first a telephone or cable customer and then become an Internet customer.

It seems obvious that whereas a generational shift will make Internet access almost universal in another 20 years, the same probably will not be said for cable TV, which may well peak and decline because there will be other ways to get television, for example, through mobile/cell phone ("mobile TV"), television on PC, laptops, PDAs, and so forth. That is the disruptive nature of the Internet, which threatens telephone companies, cable companies, and TV broadcasters alike. The result is that each of these industries is trying to enter the other's market. As such, cable TV is the very heart of the U.S. broadband industry, even though broadband is what will probably end up eating into cable's business. Telephone companies and Internet providers are also trying to find ways to enter the television business while VoIP is cutting into their voice business. While waiting for the intergenerational boom or bust, which is cyclical, each industry is building out to maximize the revenue from its existing subscriber base. Cable TV companies do this by offering pay-per-view, digital cable, and video-on-demand. Telephone companies are starting to do the same. However, Internet companies have a slightly different task, and that is finding ways to connect more devices and more device types to their networks. That is because the Internet of today offers nothing more than the connection and the bandwidth. The value of the Internet is increased solely by the number of connections to it. In a nutshell, if the Internet industry is close to maxing out with connecting laptops, it is logical to start connecting non–PC-type devices (see Fig. 1.2). This is most probably the number one reason and motivation for the start of the concept of connected home and home networking.

Because in the past few years many TV shows and movies have suddenly been made available over the Internet, it has driven more types of devices to be connected to networks, which in turn will increase the business value of a network to its nominal owners. Figure 1.3 shows multiple devices connected to broadband. The real value here is to acknowledge that network unification is just a technical urge. The actual value achieved will drive all the existing networks into a single technology with wired and wireless varieties. This is a tectonic shift; it is slow but inevitable and

Home Security Systems	Dual-Mode Phone	PC and Laptop	Utility Meters	Digital TV / VoD	Video Phone	HD/ PVR	Hand-held PDA	802.11 Handset	MP3 Music	Gaming

The "Connected Home" is Shaping
the Broadband Vision

FIGURE 1.3   More devices are being connected to broadband.

also irresistible. Of course, this vision of single technology is subject to national and local politics and short-term business advantages, but the trends are clear:

- All networks will eventually be collapsed, converged, merged, or subsumed into the Internet.
- Sharing the increased value of a single larger network will be worth more for information technology (IT) and information communication technology (ICT) companies than the incremental revenue from services running through different networks. Some of these cost savings could be passed onto the consumer, which will further fuel this trend.
- The above trends will force a change in the network architecture. The client–server model has dominated the Internet for most of its existence, which is also the main cause of the scaling problem today. A question the reader may ask is how big the data center should be? before realizing that no data center is big enough for some applications, especially in the long run. In order to provide hundreds of millions of simultaneous unicast high-definition television (HDTV), data centers would need to be placed close to the consumer, which is certainly going to be very expensive. Surely, this is no way to make money. In other words, only server-to-server and peer-to-peer architectures make sense in the end. This is pretty much the only way the system can scale high enough to functionally improve and then replace today's TV.

1.2   TECHNOLOGY ADOPTION TRENDS

What drove the entire semiconductor industry for the past 35 years was Moore's law. Moore's original statement can be found in his publication "Cramming more components onto integrated circuits" (*Electronic Magazine*, 19 April 1965):

> The complexity for minimum component costs has increased at a rate of roughly a factor of two per year . . . . Certainly over the short term this rate can be expected to continue, if not to increase. Over the longer term, the rate of increase is a bit more uncertain, although there is no reason to believe it will not remain nearly constant for at least 10 years. That means by 1975, the number of components per integrated circuit for minimum cost will be 65,000. I believe that such a large circuit can be built on a single wafer.

Applying Moore's law, as shown in Figure 1.4, to the networking and communication industries meant doubling the price-to-performance ratio every 18 months, which drove the IT industry. The same is expected to happen for the bandwidth-to-price ratio in networking, communication, and home networking industries.

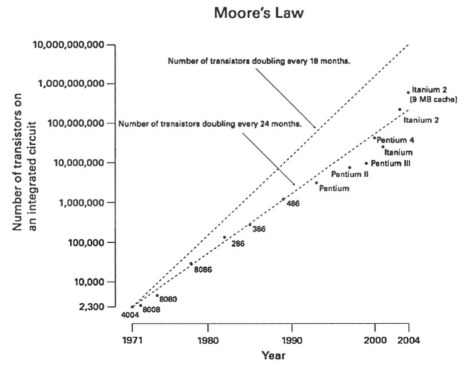

FIGURE 1.4   Growth of transistor counts for Intel processors (dots) and Moore's law (upper line = 18 months; lower line = 24 months). *Source*: Wikipedia [1].

High-definition entertainment, online gaming, and so forth, will load the networks according to Moore's law and drive this industry for at least the next decade or two.

During the early days of the Internet, universities were the first to embrace Internet services, such as e-mail and newsgroups. Financial industries and then general businesses were quick to follow in order to automate and improve their business processes, transactions, communications, and thereby reduce the overall operational costs. Service providers (SPs) were the last due to the slow deregulation of the telecommunications industry (unbundling of local loop), time to roll out massive network infrastructure, and high cost for bridging the last mile (see Fig. 1.5 for early adopters of Internet). These days, there is a paradigm shift in the adoption of innovative Internet services such as online gaming, social networks, file sharing, music/video sharing, and so forth. Universities are once again the early adopters of new services. Consumers then follow quickly, and then service providers offer those services and build their business model around those services. Finally, businesses adopt these services as they begin to understand the impact of these services on their business itself, and how these new trends, behavior, and expressions could be used as a vehicle for corporate messaging, marketing, product positioning, and attracting top talent. Thus, consumer electronics is driving the networking needs as never before, and this trend is only going to accelerate with the transport of all services and applications over the Internet Protocol (IP).

As consumers demand faster broadband communications and entertainment services, demand for wired homes increases particularly at the higher end of the housing market. Real-estate developers are increasingly turning to optical technology with several fiber-to-the-home (FTTH) vendors working closely with master planned communities (MPCs). MPCs are residentially focused real-estate developments in which builders plan an entire community around shared services and amenities that attract buyers and renters alike. MPCs are leveraging more and more FTTH to provide premium services, which is mainly driven by the decline in the optical equipment costs. The benefit of FTTH is its bandwidth capacity, which is much more than needed for voice and data services. SPs could benefit from this increased bandwidth capacity by rolling out IPTV and IP video.

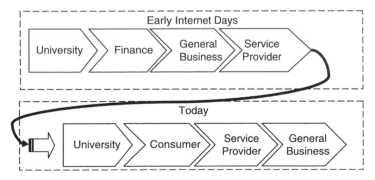

FIGURE 1.5 Early adopters on Internet: past and present.

1.3 SOCIAL NETWORK

What is a social network? Is this a trend or a passing fad? What does it mean for communications, computer, consumer electronics, and media/content industries and how does it influence home networking? A social network is a human empowered network. A social network service is a platform specifically focused on the building and verifying of social networks for whatever purpose. Blended networking is an approach to social networking that combines both offline elements (face-to-face events) and online elements. Many social networking services are also blog-hosting services. As of 2005, there are more than 300 social networking Web sites.

Social networking is a platform where people from all walks of life come together to express themselves by means of sharing videos, music, pictures, content, and so forth, and it provides the ability to collaborate using peer-to-peer applications and services. It provides a sense of virtually connecting people and platforms to share similar interests. This is a megatrend, and we have only witnessed the tip of the iceberg. The first social networking Web site was Classmates, which began in 1995. It was widely used in virtual communities. The popularity of these sites grew rapidly. By 2005, MySpace was getting more page views than Google. MySpace recently reported (mid-2006) that it has more than 100 million members and is adding half a million new users every week. Google recently bought $900 million worth of advertisement on MySpace. Bebo is one of the fastest growing social networks in the United Kingdom. Major telecommunications operators and businesses are just beginning to understand how these social networking services could have major societal impact on their service offerings, products, price, time-to market, and so forth.

According to Reuters, YouTube is working to build a library of every music video ever created. Even media and label giants plan to work with YouTube on this project. Media/label/record companies are obviously interested in legitimate use cases, but they are more interested in finding out where this trend is going and how quickly they can adapt to it. Google Video also plays into this space. Major portal vendors like Microsoft launched Windows Live Spaces, which initially received a lot of attention. The strong MSN brand and its broad product appeal marries Instant Message (IM), social networking, gadgets, blogging, and mashup under one umbrella. A blog is a user-generated Web site where entries are made in journal style and displaced in a reverse chronological order [2]. A mashup is a Web site or an application that combines content from more than one source into an integrated experience [3]. As of June 2006, Windows Live accounts for 30 million users, second only to MySpace.

XuQa (the name is derived from "hookah") is one of the me-too social networks focused on games such as poker with roughly 1 million users. It involves users participating in one big game. XuQa users play for peanuts by adding friends, uploading pictures, watching advertisements, and so forth. XuQa wants its users to sit through the advertisements and sign up for affiliated programs. Active users climb through 10 levels in the game for the ultimate prize of exposure on the XuQa home page and $1000 cash. AirG, a Vancouver-based mobile social network, is very promising. It is changing the way young people share and collaborate on content (e.g., AirG, which hit the milestone of 10 million unique users, grew from 7 million users in just a few months). What makes mobile social networks like AirG so special? AirG has no PC-based network and works exclusively through a mobile phone interface unlike MySpace and Facebook, which both work with the PC and the handset. This indicates that mobility and the ability to do everything a PC can do while on the move at any time is not hype but soon to be the reality for the young generation. Flickr is a unique example of human networking. It is an online picture-sharing site where people can place comments and also critique them. People are now more than ever able to share their thoughts, comments, and ideas. Corporations are seeking these individual talents for their businesses and for their marketing campaigns. Social networks have opened doors to people from anywhere in the world to participate and expose their talents on a global basis.

1.3.1 Business Applications

Social networks connect people with all different types of interests, and one area that is expanding in the use of these networks is the corporate environment. Businesses are beginning to use social networks as a means to connect employees and to help employees build profiles. This makes them searchable and connected with other business professionals. One example of a business social network is LinkedIn, a network that connects businesses by industry, functions, geography, and areas of interest. Networks are usually free for businesses or at a low cost; this is beneficial for entrepreneurs and small businesses looking to expand their contact base. These networks act as a customer relationship management tool for companies selling products and services. Companies can also use social networks for advertising in the form of banners and text advertisements. Because businesses are expanding globally, social networks make it easier to keep in touch with other contacts around the world.

1.4 CONSUMER TRENDS

The market in general is at an inflection point where enterprises and businesses are no longer in power but rather the end-users are (empowered human). Businesses harness the end-user energy to improve productivity. Some estimates claim that 40% to 50% of productivity can come from this kind of collaboration.

Success in the consumer market and home networking is no different. It is not any more about technology but rather about emotional connection that people are going to make with brands and products. Take for an example Apple and iPOD. Consumers expect applications and services as shown in Figure 1.6, all in real time. For the consumer, it is about seamless delivery of applications, services, and entertainment anytime, anyplace, through any media player as shown in Figure 1.7. This implies convergence of applications, services, networks, and industries such as telecom, cellular, computing, media broadcasting, and so forth, with terminals.

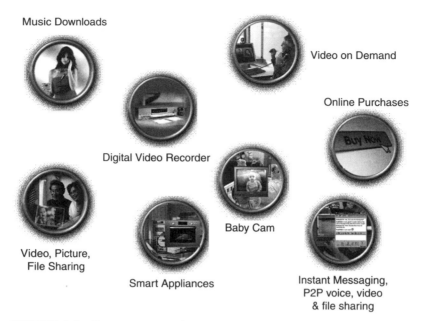

FIGURE 1.6   Consumer expectation of applications and services all at real time.

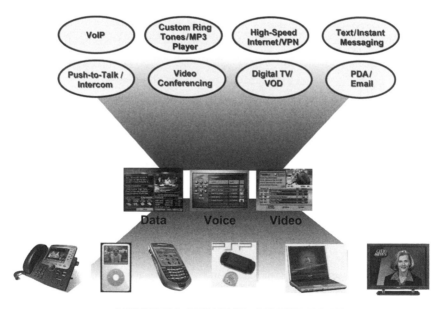

AT WORK, AT HOME, ON THE ROAD

FIGURE 1.7   Many services to many screens.

1.5 LIVING IN REAL TIME

Doing things in real time over the net provides the real added value of the networked home, which is much more than simply accessing content stored locally in the home. Nowadays, everything resides somewhere on the network; living in real time means having access to network at all times and having some control capabilities for home monitoring, appliances, and other devices from remote locations. For it to become reality, convergence of information, commerce, and communications services is a must. Real-time information would include services such as news, education, search engine, health care, travel, TV, and so forth. Real-time commerce would mean online payment and online banking services. This will enable services such as online reservation and payment for movies, games, music, ring tones, entertainment, and ordering of goods and services. Communication services include telephone, video telephony, instant messaging, e-mail, multimedia services, and so forth.

1.6 CONFLUENCE OF EVENTS

The mobile industry is huge with well over 2.3 billion subscribers and growing at more than 800 thousand subscribers per day. The majority of the growth is coming from the emerging economies with high populations, such as Brazil, Russia, India, China, Bangladesh, and so forth. With the increasing competition among fixed network operators and mobile operators, the voice tariffs are under huge pressure to decline.

In the meantime, explosion of broadband to home happened over the same period, and with more than 200 million connections worldwide and growing, free peer-to-peer Voice-over-IP (VoIP) and Session Initiation Protocol (SIP)-based voice services such as Skype became increasingly popular among residential users, teenagers, and students. Moving forward, VoIP is expected to take an increasing bite off the voice call market and will replace time- and distance-based charging with service-based flat tariff. Consequently, fixed telecom and cable operators have begun to expand their services offerings to triple play, namely, voice, video, and data (Internet connectivity) to combat declining voice revenue. During the same period, standardization of IEEE 802.11a/b/g standards and certification of Wi-Fi product granted vendor interoperability, which resulted in increased market competition and drove price down for Wi-Fi chipsets and access points (APs). Soon, embedded wireless local area network (WLAN) chipsets became the mainstream feature just like infrared radio (IR) in laptops.

Development of new service offerings, particularly multimedia applications and the notion of converged networks based on IP, have forced the equipment manufacturers to consider multiple wireless connectivity solutions in products such as Bluetooth, Wi-Fi, IR, and so forth, in order to enhance consumer experience and meet their rising expectations. Figure 1.8 illustrates the confluence of events driving fixed-mobile convergence (FMC). Dual-mode phones and devices with Wi-Fi capabilities will leverage all these trends to connect to the mobile network and to Wi-Fi APs, which act as IP gateways between broadband IP-based fixed network and mobile networks. The device uses client software to provide soft switching between mobile and fixed networks for continuity of voice call and for call handoff.

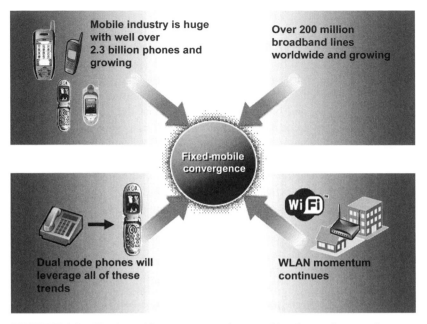

FIGURE 1.8   Fixed–mobile convergence is created by the confluence of events.

For now, two approaches to client and gateway functionality have emerged, namely Unlicensed Mobile Access (UMA) and SIP. Ultimately, these approaches will converge into IP Multimedia Subsystem (IMS).

1.7   APPLICATION AND SERVICE CONVERGENCE

Application convergence combines electronic subsystems that meld voice, data, and video into all sorts of electronic equipment, and with the availability of nanometer process technology, silicon IP vendors can use System-on-Chip (SoC) to develop, design, and build solutions. Application convergence means that different technologies combine for one application such that various traffic types such as video, communication, and computing are in the same SoC design. Figure 1.9 depicts convergence of applications, services, and networks. An example is when computer-based applications like e-mail and customer relation management converge with communications applications like telephone calls and voice mail.

The goal of service convergence is the fast delivery of voice, video, and data services with full mobility in an end-to-end secure architecture. Service mobility implies service continuity across network domains and devices. Wired, wireless, and mobile service providers are jockeying to own the customer. This will have an effect on the value chain for service delivery of entertainment, information, and communication. They understand that their ability to provide voice, video, and data capabilities for converged services will increase their customer loyalty, stickiness, and draw new

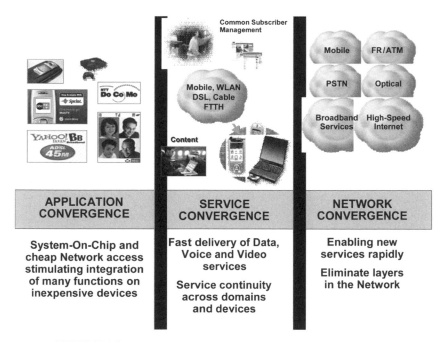

FIGURE 1.9 Convergence of applications, services, and networks.

subscribers. Service convergence is not about quadruple play, which is voice, video, Internet, and mobility, but rather it is the ability to create new and innovative services based on the convergence of voice services, Internet services, and video services with mobility [4]. First, the architecture of these services will be loosely coupled where end devices will interwork to provide a converged service and later it will be a tightly coupled architecture where network and/or middleware will be intelligent to interwork with other services and peripheral networked devices to provide an even richer converged service user experience. FMC is a great example of service convergence. Another example of converged service is remote home monitoring and management service (see Section 1.10.4).

VoIP is currently being promoted as a replacement to traditional phone lines. Calls can be made on the Internet using a VoIP service provider and standard computer audio systems, a VoIP phone, or another similar multimedia device. Alternatively, some service providers support VoIP through ordinary telephones that use special adapters to connect to a home computer network. Converged messaging services such as Unified Communications (UC) services are expected to improve personal productivity and hence positively contribute toward the enhancement of human life. The first step of UC application will bring voice mail, e-mail, and faxes into a common inbox where they can be deleted, answered, forwarded, or saved. The next step for UC is integration of real-time presence application in order to reduce the number of unnecessary calls. It is not far when UC application will work with other wired and wireless peripheral networked devices to provide easier, faster, and better user experience; for example, UC might include videoconferencing, file sharing, white board, a standard desktop/laptop and telephone feature, rather than something that needs to be specially scheduled. Just clicking on an icon brings additional participants into the discussion with other home-networked devices.

Service convergence will not only improve the home user experience. Service providers are coming out with hosted solutions that harness UC technology to make contact centers more versatile than ever. These hosted solutions make it possible for contact centers to be onshore, near-shore, or offshore without heavy investment in infrastructure. As a result, the architecture of the center is highly flexible based on the needs and/or culture of the operation (i.e., they can be centralized, distributed, or home based). Using UC and the hosted application model, for instance, the call center can add home-based workers easily or ramp up existing operations at the drop of a hat. Thus, companies can extend their facilities across multiple time zones and locations and change their structure to fit seasonal loads.

The real benefits of UC for businesses is physical virtualization of call center operations that span continents via distributed contact centers and home-based agents. It may not make economic sense in certain geographical regions to set up full-fledged and dedicated contact centers to directly service the emerging middle class. However, the combination of UC and hosted call center solutions opens up new opportunities and new markets. Virtual contact centers can be deployed to service any language without the necessity of investing in infrastructure for each country. Some countries perhaps can have centers specific to their region, whereas others can cater to several languages using multilingual agents or via a network of home-based customer service representatives in the countries required. Another good reason for the marriage of UC and hosted solution is that in a global and virtual operation, it is desirable to have the same technology environment in every location. This makes it easier to manage globally and to keep the costs down. Further, the customer gains a uniform experience no matter where they call in the world.

1.8 NETWORK CONVERGENCE AND REGULATIONS

Mobile networks, fixed networks, and unbundled local loops all had a major impact on network convergence. Service providers competing to acquire new customers are faced with high subscriber churn rates, voice traffic moving from fixed to mobile networks. For any given period of time, the number of participants who discontinue their use of a service divided by the average number of total participants is known as the churn rate. Churn rate provides insight into the growth or decline of the subscriber base as well as the average length of participation in the service. Furthermore, the rapidly declining voice tariffs are forcing the traditional network operators to reduce the total cost of ownership and operational expenditure by offering innovative services over a single IP-based network. Dual-mode phones are able to connect to mobile networks and Wi-Fi APs at home, public Wi-Fi hotspots, airports, hotels, work, and so forth, which are connected to an IP gateway between fixed and mobile networks. The purpose of this architecture is to enable an active call on any dual-mode phone to be switched (handover) back and forth between fixed broadband and mobile networks depending on the best available network to reduce usage of licensed spectrum and to enable users to call using a single device at the lowest cost and to maintain the call connectivity.

In many countries today, IP services such as VoIP, IPTV, and so forth, are treated as information rather than as telecommunication services. In some countries, national policymakers and regulatory authorities are revising licensing frameworks to make them more flexible and propose a generic or converged license for all forms of telecommunication services regardless of the underlying technology deployed or service offered. Some countries are exploring the possibility of decoupling the network operations license from service provision, whereas other countries prefer a liberalized communication environment.

1.9   TERMINAL CONVERGENCE

Development of new applications and services such as IP multimedia are forcing terminal manufacturers to consider including multiple wireless interfaces in products to provide even richer customer experience. Terminal vendors are challenged with the difficult task of balancing functionalities of devices with cost, power consumption, and the customer's ergonomic expectations. This requires the terminal manufacturers to provide product differentiation in order to be competitive in the marketplace. This pressure will ultimately be passed on to silicon vendors for all converged devices and

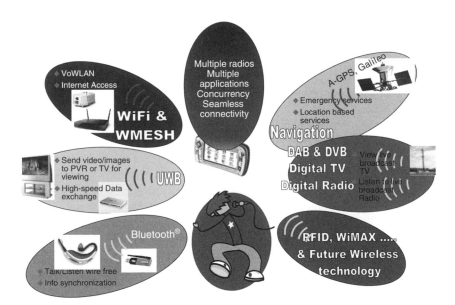

FIGURE 1.10   Convergence possibilities: new applications, new services, single terminal, and so forth.

terminals for home. A good example of this phenomenon is the mobile terminal market, which is subjected to most intense pressure and is all about unit volume shipments. With time, more and more radio technologies will find their way to smart devices, and handheld terminals will lead the way by implementing, for example, WiMAX (2.5 GHz) and ZigBee (2.4 GHz) for home automation applications. There already are six radios operating in 10 frequency bands within a single device such as GSM/GPRS/EDGE (850/900/1800/1900 MHz), WCDMA (1.9–2.1 GHz), WLAN (2.4 GHz), GPS (1.5 GHz), Bluetooth (2.4 GHz), and FM Radio (88–108 MHz). Figure 1.10 illustrates concurrent seamless connectivity from the convergence of multiple radios and applications.

1.10 HOME NETWORKING

Convergence of applications, services, and devices over IP-based networks is resulting in a converged world leading to the evolutions of new business models. Network adaptability will be the key criteria to help meet the accelerating changes and make network responsive to the diversified needs and requirements. In addition, the businesses will need to be more agile than ever before.

For business of the future to be agile, the questions that need to be answered are (1) Where should intelligence reside in order to deliver next-generation services—in networks, in devices (software), in applications, or all of the above? (2) What will be the next-generation platform for service creation? Most likely, it will be the battle between different approaches and visions from different industry segments, such as network giant (Cisco), media player giants (Apple, Sony, Microsoft), mobile/ wireless networking and devices giants (Nokia, Ericsson, Motorola, Samsung), and applications/services giants (Google, Yahoo, Microsoft). These companies will push their architectural models for business acceptance, but ultimately, the platform of choice will be the one that will be open for businesses and developers to build their value-added services on.

Home networking may be categorized under four main segments, namely, home computing, home network entertainment, home communications, and home monitoring and management, as shown in Figure 1.11. The core of home computing functionalities includes interworking between indoor (local area network) and outdoor (wide area network) networks and interconnecting multiple devices at home. Home network entertainment includes all networked consumer electronic devices such as X-Box, PS3, networked DVR, camcorder, TV, and so forth. Home communication includes voice and video telephony, Wi-Fi phones, dual-mode phones, and so forth. Finally, home monitoring and management includes wireless remote control, remote surveillance and home systems management, and so forth.

There is a major industrial initiative called Digital Living Networking Alliance (DLNA) to develop innovative new technology and to guarantee interoperability by leading consumer electronic, PC, and mobile handset companies, which allows users to easily acquire, store, and access digital content from almost anywhere in

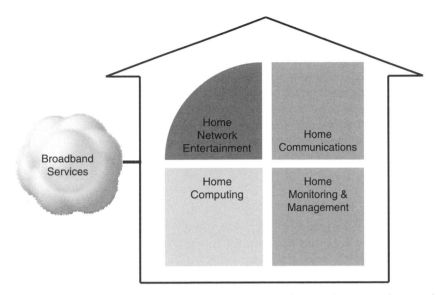

FIGURE 1.11   Home networking segmented into personal computing, network entertainment, communications, and monitoring and management.

the home. It enables users to effortlessly manage, view, print, and share their content. For example, on a DLNA home network, it is possible to access a home video from home digital video recorder and watch it on a PC anywhere in the home [5].

1.10.1   Home Computing

The basic requirements on home networking from the computing perspective are:

- Interworking and distribution of high bandwidth to multiple devices in the home
- Interworking between home IP gateway and wide area network (WAN) connection
- Securing in-home communication and via WAN connection
- Wireless enhancements for range and speed
- Wired and wireless home network media
- Distributed file sharing

New higher speed access with freedom of movement in the home while maintaining security and privacy demands technologies that are capable of delivering secure high-speed quality and high-definition entertainment. Some of these are as follows:

- Reliable wireless bandwidth over 100 Mbps (IEEE 802.11n). Higher bandwidth and longer-range wireless will enable home networks that can distribute high-bandwidth video content in and around the home.

- No new wires technologies such as
  - High-speed networking over existing coaxial cable (MoCA).
  - High-speed networking over existing telephone line (HomePNA).
  - High-speed networking over existing powerlines (HomePlug AV). HomePlug AV enables networked entertainment and Broadband over PowerLine (BPL) applications for home networking. HomePlug AV-based integrated circuit (IC) will allow consumers and service providers to distribute high-definition video and audio over existing in-home electrical wires, ushering in a whole new era of user-friendly entertainment connectivity.

1.10.2   Home Entertainment

Entertainment on demand is on the march with the consumer. An example of a networked platform that offers consumer sticky application around digital media is consumer place-shifting device or software. Sling Media offers a consumer device that sits next to a residential digital set-top-box and allows a user to view his or her home-pay TV programs via Internet on a PC or on a mobile. On the other hand, Orb Networks' solution is entirely software based, installed on PCs and mobiles. Orb announced a working relationship with Advanced Micro Devices (AMD), which might have important industry ramifications. Its technology will be the foundation for AMD's Live media server capabilities and part of AMD's Live entertainment PC platform. This will be the first time the PC industry puts ready-to-go in-built capabilities such as anytime, anywhere media streaming on PC into the hands of consumers, allowing them indoor and outdoor content delivery.

Content storage for home is a fragmented market, and there is little demand for consumer-networked storage. Consumers store their files and digital media on their PC/laptop hard drives, external hard drives, USB drives, and/or burn CDs/DVDs. TV content is stored on DVRs and/or on PC with TV card. The home-networked content storage market is still underdeveloped but with consumer demand toward producing personal video content, sharing files, P2P video file transfer, IPTV, and so forth. Consumers are starting to see a large collection of their digital media files stored across multiple devices. This will generate demand for low-cost networked storage devices that can be accessed through various media players through different mediums (e.g., TV, PC, mobile phone, iPOD, MP3 player).

1.10.3   Home Communications

Communication in-home involves the following:

- Integrated communications for fixed and mobile: FMC is a great example for integrated fixed and mobile communications discussed earlier in this chapter. Another great initiative to promote FMC is the Fixed-Mobile Convergence Alliance (FMCA), which is a global alliance of telecom operators whose objective is to accelerate the development of convergence of products and services [6].

- Instant communications with federated user groups: User Personal Network (PN) device will enable a user to add and remove other users from his or her federation. Users belonging to a federation will be able to share all forms of communications instantly such as voice, video, IM, and so forth, with other users within the same federation. For more on the concept of PN-to-PN communications and PN-to-federation solutions, please refer to a worldwide R&D project called "My personal Adaptive Global NET (MAGNET)" [7].
- Presence integrated into voice communications: Voice communication systems integrated with real-time presence application will provide the users with the ability to see who else is logged-on to the network at the time and if the person on the other end of the communication link is willing to receive a call. This reduces the number of unnecessary calls and improves human interactions. This capability is currently available for corporate users, and it is expected to enter the residential market as VoIP begins to dominate the home communication market.

To provide complete freedom of movement both to the user and home devices and to seamlessly connect them, wireless networking is emerging as the major new growth area of research and business opportunities. This would enable control, access, and information sharing among all the devices and much enhanced user experience.

1.10.4 Home Monitoring and Management

There is growing importance of home monitoring and management. Recently, AT&T announced its new "AT&T Home Monitor" service, which allows subscribers to monitor their homes remotely, through a PC or AT&T (formerly Cingular Wireless) mobile phone. A key perceived benefit of this service is the ability to monitor activities of children and elderly parents remotely. With the ageing population, this service is poised to take off because it helps to reduce the cost of health care. In addition, it allows remote control of home lighting and potentially other networked appliances and receives a range of alerts and reports on conditions in the home, through a variety of motion detectors and temperature sensors.

In monitoring and management, the following items have been characterized:

- Remote surveillance
- Home systems management
- Heating, ventilation and air-conditioning (HVAC) and lighting control

Service providers are motivated to offer this service because such services will help make their overall service offering attractive for subscribers. This will increase their subscriber stickiness and reduce the subscriber churn rate. Discontinuing the service would mean subscribers would lose their home monitoring service. Providers are also quite interested to offer these kinds of converged services due to the migration of appliances, sensors, and network architectures toward an all-IP–based technology.

1.11   CONNECTED HOME

In-home, there is an explosion of rich-media devices that can access different types of content blurring the lines between consumer electronics, computing technologies, and communication devices, delivering the same quality media to empowered consumers for in-home entertainment.

Even though there is blurring of media devices, there are still many networks in-home that consumers have access to such as cable, DSL, satellite, fixed phone, IP phone over cable/DSL, FTTH, mobile phone, and so forth, as shown in Figure 1.12. Consumers demand simplicity—all they want is the availability and accessibility to services preferably over a single network. They do not care about which network does what, nor do they care which service provider provides what. Therefore, the service providers are in the process of reducing network operational cost by consolidating disparate networks into a single IP-based network so that a consumer only needs a single connection to his or her home.

For consumers to have truly rich experience from a connected home, the following needs to happen:

1. Convergence of communication, computing technologies, and consumer electronics.
2. Delivery of rich media and content over a single network by service providers.

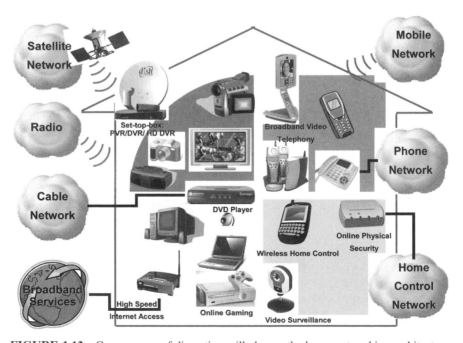

FIGURE 1.12   Convergence of disruption will change the home networking architecture.

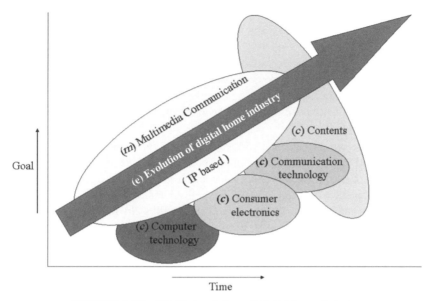

FIGURE 1.13   (E)-/(re)-volution of digital home industry.

3. Seamless interworking and sharing of content between media devices and players.
4. Simplified user interface and preferably a single lightweight device for home networking control and management.

Based on the topics discussed in this chapter such as technology adoption trends, consumer trends, and convergence of networks, applications, services, and terminals and from the section on home networking, we can derive an equation that will drive the future development of digital home industry:

$$E \propto mc^4 \qquad (1.1)$$

where E is evolution of digital home industry, m is multimedia communications, and c^4 stands for content, consumer electronics, computer technology, and communications technology. Figure 1.13 illustrates the clue to the (e)-/(re)-volution of digital home industry [8].

1.12   VISION OF THE FUTURE

Niels Bohr, the famous Danish physicist who made fundamental contributions to understanding of atomic structure and quantum mechanics, for which he received the Nobel Prize in 1922, said "Technology has advanced more in the last thirty years than in the previous two thousand. This exponential increase in advancement will only continue" [9].

Here we have what might be three new laws to guide future growth of the Internet, and they tell us a lot about what the future holds for us.

1. It is logical to assume that all devices will eventually be networked, but it will most probably start with devices that either are already intelligent (computers) or are very important to us (e.g., television, phone, and music system). Any political or business obstacles in the path of this transition are not likely to stay for long.

2. A possible way for the network to maximize its value is through server-to-server and peer-to-peer networking. Therefore, if ISPs do not want consumers to share or re-sell their Internet connection, it is just because they do not yet realize that such sharing is really to their advantage. In time, they will come to see it or they will slowly disappear. If ISPs do not like all that heavy traffic, it is just because they have not yet understood that in the long-term, they have to find ways to monetize the connections to be successful.

3. As we look into the future, the rate of change appears to be accelerating. Devices, systems, services, and applications capabilities are defined by software, which is changing much faster than hardware, mainly because the services today seem to last for a much shorter period than they used to when targeted toward the consumer market. Future generations will drive the demand for new and advanced information services. The future of home networking will heavily revolve around this industry's ability to improve the way we entertain, communicate, educate, and socialize in a secure way [10].

For the industry, this means that:

1. Market transitions will be customer driven.
2. Technology architecture with an economically viable business model will be required.
3. Networks and systems complexities must be hidden from the end-users.

1.13 BRIEF OVERVIEW OF THE BOOK

This book on home networking has been organized to delve into the important enabling technologies to realize a home communications environment that improves the user's experience, quality of life, and ease of use. This has meant covering other topics than only just those on networking. In addition to presenting the material in an easy to understand tutorial manner, the chapters also include the latest research results and technical advances on the various topics. To make all this happen, a holistic approach has been adopted to develop an intelligent wireless home ecosystem. A list of topics that are covered in this book is provided below.

- Networked home use case scenarios
- Media formats and interoperability

- Media description and distribution in content home networks
- Mobile device connectivity in home networks
- Generic access network toward fixed-mobile convergence
- Secured personal wireless networks: home extended to anywhere
- Usable security in smart homes
- Multimedia content protection techniques in consumer networks
- Device and service discovery in home networks
- Small, cheap devices for wireless sensor networks
- Spotting: a new application of wireless sensor networks in the home

The book has been organized as follows. First, we have presented some background and the key use case scenarios in this chapter to set the scope and give a perspective on the needed enabling technologies to realize an intelligent networked home. Because content distribution is one of the main functions performed by a home communication system, we therefore focus in Chapters 2 and 3 on the media descriptions, media formats, their interoperability, and distribution. In Chapter 4, we present how the mobile devices may be connected in the home network system, and what requirements do they impose on such a network both from the local connectivity and remote connectivity perspectives. Chapter 5 focuses on the Unlicensed Mobile Access (UMA) and Generic Access Network (GAN) protocols to enable fixed–mobile convergence in the home. In a virtual sense, the home network could be geographically distributed provided necessary security and authentication mechanisms have been built in. This is the topic of Chapter 6. As users become reliant on mobile devices to handle sensitive information in the home, requiring the home infrastructure to become increasingly wireless, there is an urgent need to provide credible solutions to security and privacy issues. This important topic is addressed in Chapter 7. A connected home will consist of multiple devices and services, which will require that they be supported by some discovery mechanisms. This topic is covered in Chapter 8. Chapter 9 tackles the issue of digital rights management in the access and sharing of the multimedia content in consumer networks, such as in a home environment. It is anticipated that the home of the future will use all kinds of sensors for monitoring, control, and data fusion. Chapters 10 and 11 delve into those issues and provide some innovative solutions and applications.

One or more subject matter experts have addressed each topic in a detailed yet understandable manner. Because the authors have adequately addressed the issues, solutions, and the cross-layer functions, we feel it would be redundant to repeat them here.

1.14   CONCLUSIONS

This chapter presented a broad view of consumer, technology, and industrial trends and interplay between them. It briefly described the phenomenon of living in real

time and the confluence of events. Further, it discussed the impact of social networking, convergence of services, applications, networks, and terminals. Finally, the
development of the connected home and a vision for future was discussed. In
summary, the key points are as follows.

Moore's law drove the semiconductor industry for the past 35 years, and applying
this law to home networking and communications would mean doubling the
bandwidth-to-price ratio every 18 months. As Niels Bohr, physicist Nobel Prize
winner, said, "Technology has advanced more in the last thirty years than in the previous two thousand. This exponential increase in advancement will only continue."
Home networking development will be driven by a new law $E \propto mc^4$, where E
equates with evolution of digital home industry, m for multimedia communications
and c^4 for content, consumer electronics, computer technology, and communications
technology. Based on this new law for the digital consumer market, we can continue
to expect major innovations in products and services for the years to come.

Proliferation of IP-enabled devices and applications for home will continue for
the near future. Focus is shifting from PC-centric devices and applications (with
already high Internet penetration) to non–PC-centric devices and applications.
These devices range from networked TV, networked DVR (IP-enabled set-top-
box), online gaming, networked camera/camcorder, networked video surveillance,
VoIP phones, dual-mode mobile phones, networked appliances, and various types
of sensors.

Place-shifting technology will revolutionize the consumer experience and expectations. It currently has the ability to deliver home TV content and manage home-
networked DVR from any PC-centric device and mobile phones from any place,
any where and any time. This technology is soon expected to enter mainstream PC
and mobile entertainment platforms with support from chip manufacturers and
service providers.

Explosion of social networks, mashups, and blogs with online capabilities of
photos and video sharing is a trend that is here to stay with us for a long time.
Home networking appliance and device vendors will try to leverage this behavior
and turn it into service and business benefits. IP-based fixed and mobile peer-to-
peer services such as voice, video telephony, and gaming are poised to take off
in a big way. This is a new trend, but globally there are still some hurdles and
regulatory issues.

Fixed Mobile Convergence (FMC) leading toward IP Multimedia Subsystems
(IMS) is currently the best industry-wide, universally accepted architecture across
fixed, mobile, and cable vendors and operators leading them toward triple, quadruple,
and multiplay solutions. This is a major accelerator for IMS technology.

The future of home networking will heavily revolve around industry's ability to
improve the way we entertain, communicate, educate, and socialize in a secure
way keeping in mind that (a) market transitions will be customer driven, (b) solutions
will be economically viable, and (c) network and system complexities must be hidden
from the end-users because simplicity of the user interface is key to mass acceptance
by consumers.

REFERENCES

1. Moore's Law. Available at http://en.wikipedia.org/wiki/Moore's_law.
2. Blog. Available at http://en.wikipedia.org/wiki/Blog.
3. Mashup. Available at http://en.wikipedia.org/wiki/Mashup.
4. *Wireless Personal Communications*. Special Issue, May 2006, Vol. 37, Nos. 3–4.
5. Digital Living Network Alliance. Available at http://www.dlna.org/en/consumer/home.
6. Fixed-Mobile Convergence Alliance. Available at http://www.thefmca.com/.
7. My personal Adaptive Global NET (MAGNET). Available at http://www.ist-magnet.org/.
8. S. Dixit and R. Prasad. *Wireless IP and Building the Wireless Internet*. Artech House, Norwood, MA, 2003.
9. Niels Bohr. Available at http://en.wikipedia.org/wiki/Niels_Bohr.
10. Mahbubul Alam et al., "Wireless Next Generation: 4G?," pages 305–316, In *Multiaccess Mobility and Teletraffic for Wireless Communications*, Vol. 5. Kluwer Academic Publishers, Norwell, MA, 2000.

2

MEDIA FORMAT INTEROPERABILITY

ANTHONY VETRO

Digital media has become an integral part of our everyday lives, and consumers have access to content through a wide array of media services. This content may be transmitted over dedicated broadband networks, wireless networks, or the Internet, and it is received on a variety of different devices ranging from set-top boxes and television sets to portable media players, computers, and cell phones. While interoperability does exist within specific application domains and media services, it is lacking among the broad range of diverse multimedia devices that could be found today. Part of the problem is that the content is compressed in a variety of different media formats making it difficult to transfer and consume content on different devices. This chapter introduces the media format interoperability problem in more detail by specifically discussing a subset of popular media formats and devices in use today. Several technologies that aim to solve the interoperability problem are then presented including the role of metadata formats and media adaptation technology.

2.1 BACKGROUND

Content enters the home through a variety of ways. Broadcast services are provided through cable, satellite, and terrestrial transmission; Internet content is accessible through broadband connections via DSL or cable; and CDs and DVDs with the latest songs, movies, and television shows are available for purchase or rental. Standards provide a means for interoperability to be achieved, and within specific application domains and services, there are no serious interoperability problems that the consumer would need to be concerned with. For instance, the DVD you

Technologies for Home Networking. Edited by Sudhir Dixit and Ramjee Prasad
Copyright © 2008 John Wiley & Sons, Inc.

just bought will not have any problem playing on your DVD player; the set-top box from your cable provider decodes and displays the compressed streams being sent to you through its service; and the song that you just downloaded from the Internet will play on your PC with the right software installed. However, with the growing demand for content portability (i.e., being able to experience content on any device), there is a distinct lack of interoperability.

As the number of networks, types of devices, and content representation formats increase, interoperability between different systems and different networks is becoming more difficult. The aim is to provide a seamless interaction between content that is authored for one purpose and consumed in a different way. Figure 2.1 attempts to illustrate this problem. On one side, we have a variety of content sources and on the other we have a set of connected devices in the home. The media formats and network i nterfaces between these devices are not uniform, thereby making connectivity and media portability a challenging task.

The rest of this chapter is organized as follows. In the next section, the various media formats that exist today will be covered, including audiovisual compression formats as well as some common transport and file formats. In the following two sections, related technologies that help bridge the gap between the different media formats, and networks that they are connected to, are discussed. The first related technology is metadata, which is useful for describing not only the content itself but also the distribution and consumption environment as well. In dynamic and heterogeneous networking environments, an essential first step is to understand the mismatch and steps that need to be taken to achieve interoperability. The second related technology is media adaptation, which is the process to convert the media from its source format into a format that could be consumed on a target device. In an effort to minimize the adaptations required in a home network, an approach that defines sets of mandatory

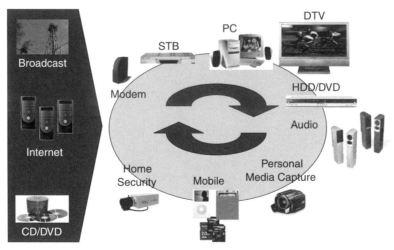

FIGURE 2.1 Illustration of several content sources including broadcast, Internet, and CD/DVD, as well as connectivity in the home among a diverse set of media devices.

media formats for a defined set of devices classes is also discussed. We then conclude the chapter with an example of media format interoperability and provide a summary of the topics covered.

2.2 MEDIA FORMATS

This section provides an overview of some common image, video, audio, and system layer formats that are supported in home network devices and discusses key applications domains for each format. We also present the concept of profiles and levels, which define interoperable conformance points for specific coding formats.

2.2.1 Image and Video Formats

Image and video coding play an important role in bridging the gap between large amounts of visual data and limited bandwidth networks for video distribution. During the past two decades, a number of image and video coding standards have been developed to satisfy industry needs. These standards have been developed by two major standards organizations: International Organization for Standardization/ International Electrotechnical Commission (ISO/IEC) and the International Telecommunication Union (ITU).

The Joint Photographic Experts Group (JPEG) has produced well-known standards such as JPEG [1] and JPEG 2000 [2]. A key difference between the two standards is that JPEG is based on the Discrete Cosine Transform (DCT), whereas JPEG 2000 is based on the Discrete Wavelet Transform (DWT). JPEG 2000 also supports higher compression efficiency, scalability, and a variety of other features such as error-resilience. JPEG is a ubiquitous image format on the Web and also the primary format used in digital cameras today, and JPEG 2000 is beginning to find applications in video surveillance and digital cinema. In addition to these standardized formats, there exist a host of other popular image formats, such as Graphics Interchange Format (GIF) (which utilizes a variation of run-length coding known as Lempel–Ziv–Welch coding), Portable Network Graphics (PNG), Tagged Image File Format (TIFF), and so on.

On the video side, the Moving Picture Experts Group (MPEG) of ISO/IEC and the Video Coding Experts Group (VCEG) study group of ITU-T have also produced a series of successful video coding standards, many of which have been jointly developed. H.261 was completed in 1990 by the ITU-T and is mainly used for Integrated Services Digital Network (ISDN) video conferencing [3]. MPEG-1 was completed in 1992 under ISO/IEC with the target application being digital storage media on CD-ROM, at bit rates up to 1.5 Mbit/s [4]. MPEG-2, which is also referred to as H.262, was completed in 1994 by both ISO/IEC and ITU-T [5]. This standard is an extension of MPEG-1 and allows for greater input format flexibility and higher data rates for both high-definition television (HDTV) and standard-definition television (SDTV). MPEG-2 is used as the video format in most digital television (DTV) systems around the world and is also the format used for DVD. H.263 was completed in 1996 by the ITU-T and was largely based on the H.261 framework [6], and

TABLE 2.1   Summary of Select Image and Video Coding Formats

Name	Major Features
JPEG	DCT-based still-image coding format used for Web pages and digital cameras.
JPEG-2000	DWT-based still-image coding format with improved coding efficiency. Used for surveillance and digital cinema applications.
H.261	Video coding format developed primarily for video conferencing applications with bit-rates in the range of 64 kbit/s to 1.92 Mbit/s.
MPEG-1	Video coding up to 1.5 Mbit/s. Target application was video storage on CD-ROM but also widely used on the Internet.
MPEG-2 (H.262)	Extension of MPEG-1 coding for higher quality applications such as DTV/DVD. Support for higher bit-rates and input picture resolutions. Most extensively used video coding format.
H.263	Designed for very low bit-rate coding, below 64 kbit/s.
MPEG-4 Part 2	Includes tools to support content-based coding and low bit-rates coding. Used for mobile video and streaming applications.
H.264/AVC (MPEG-4 Part 10)	Significant improvements in coding efficiency over MPEG-2 and MPEG-4 Part 2. Wide application for DTV and HD-DVD.
VC-1	Coding performance close to H.264/AVC and also targets a wide number of DTV/DVD applications.

MPEG-4 Part 2 was completed in 2000 by ISO/IEC [7] and is compatible with H.263. The Simple Profile and Advanced Simple Profile of MPEG-4 Part 2 have been used for mobile application and streaming. Finally, the most recent video coding standard, H.264/AVC, which is also referred to as MPEG-4 Part 10, was developed by the Joint Video Team (JVT) of ISO/MPEG and ITU-T/VCEG [8]. H.264 greatly improves the coding performance over MPEG-2 and MPEG-4 Part 2 by achieving the same quality at approximately half the bit-rate. The target applications include broadcast television, high-definition DVD, and digital storage.

A summary of the image and video coding formats is shown in Table 2.1. Currently, the most popular video coding standards include MPEG-2, MPEG-4 Part 2 (Simple Profile and Advanced Simple Profile), and H.264/AVC (Baseline Profile). It should be noted that besides the video coding standards developed by MPEG and VCEG, there are also video coding schemes such as VC-1 (informal name of the SMPTE 421M video codec standard) developed by Microsoft and standardized by the Society of Motion Picture and Television Engineers (SMPTE) [9], as well as RealVideo, which is a proprietary format developed by RealNetworks. Such media formats are extensively used for video streaming over the Internet. In the case of VC-1, it has also been adopted into the specifications developed by the BluRay Disc and HD-DVD Forums.

2.2.2   Audio Formats

Audio coding has been central to the deployment of a wide variety of services including television and radio broadcast, as well as music distribution via CD or over the Internet.

The most popular coding formats for streaming audio on the Internet today include MP3, a standardized coding format developed by MPEG, as well as the proprietary Windows Media Audio (WMA) and RealAudio formats, which have been developed by Microsoft and RealNetworks, respectively. For high-quality applications, such as music CD, television broadcast, and DVD, the uncompressed Linear Pulse Code Modulation (LPCM) format, as well as Advanced Audio Coding (AAC) and AC-3 (also known as Dolby Digital) compression formats, have been more prevalent. A summary of select audio coding formats is shown in Table 2.2.

LPCM is an uncompressed audio format [10]. With up to 8 channels of audio at 48-kHz or 96-kHz sampling frequency and 16, 20, or 24 bits per sample, the maximum bit-rate could go up to 6.144 Mbyte/s. However, the format encoded on most CDs and DVDs use 48-kHz sampling and 16 bits per sample.

MP3 is a perceptual audio encoder that encodes the audio signal based on the characteristics of the human perception of sound [11]. In such as scheme, the parts of the audio signal that humans perceive distinctly are coded with high accuracy, and the less distinctive parts are coded less accurately. Typical MP3 bit-rates are in the range of 8 kbit/s to 320 kbit/s. The sampling frequencies defined by MPEG-1 include 32 kHz, 44.1 kHz, and 48 kHz. MP3 also works on both mono and stereo audio signals.

MPEG-2 AAC is a more efficient and flexible encoding scheme and is able to achieve perceptually transparent quality at only 64 kbit/s per channel [12]. Sampling rates range from 8 kHz up to 96 kHz and above, and bit-rates may go as high as 256 kbit/s. The standard also supports up to 48 channels, as well mono, stereo, and all common multichannel configurations such as 5.1 or 7.1. MPEG-2 AAC has been adopted by both the European and Japanese digital television broadcasting systems.

TABLE 2.2 Summary of Select Audio Coding Formats

Name	Major Features
LPCM	An uncompressed audio format used in CD and DVD. Typical sampling rate of 48 kHz and 16 bits per sample.
MPEG-1 (MP3)	Perceptual audio encoder with bit-rate in the range of 8 to 320 kbit/s. Popular audio format on the Web and in portable music players.
MPEG-2 (AAC-LC)	Perceptually transparent quality at only 64 kbit/s per channel. Support for up to 48 channels and bit-rates up to 256 kbit/s. Adopted for European and Japanese broadcast.
MPEG-4 (HE-AAC)	Improved coding efficiency. Good quality stereo at 16 to 24 kbit/s. Used for satellite radio and adopted by 3GPP.
AC-3 (Dolby Digital)	First coding format specifically designed for multichannel sound. Adopted for use in U.S. terrestrial broadcast, satellite television, DVD.
WMA	Popular format for Internet-based streaming applications. Also supported in consumer electronics devices.

MPEG-4 HE-AAC, which stands for high-efficiency AAC and is also referred to as aacPlus, improves coding efficiency with the use of a bandwidth expansion tool referred to as Spectral Band Replication (SBR) [13]. With this coding tool, it is possible to achieve good stereo quality at bit rates of 32 to 48 kbit/s. In the second version of this standard, parametric stereo coding tools are also added, which reduces the required bit-rates to around 16 to 24 kbit/s for stereo content. These coding techniques have been adopted by 3GPP for mobile music services and for satellite radio providers such as XM Radio. There are a variety of additional flavors of AAC that target low-delay, scalability, and lossless compression. Further details may be found in Ref. 14.

AC-3 is also a perceptual digital audio coding technique. AC-3 was the first coding system designed specifically for multichannel digital audio, but it also supports mono and stereo coding as well. Because an early target of this audio coding format was television broadcast in the United States, the specification of AC-3 is defined by the Advanced Television Systems Committee (ATSC) [15]. AC-3 supports sampling rates of 32, 44.1, and 48 kHz with bit-rates in the range of 64 to 640 kbit/s. Besides being adopted for terrestrial digital television broadcast in the United States, AC-3 is also widely used for satellite television systems, DVD, and gaming.

WMA and RealAudio are proprietary formats that are mostly found on the Web for streaming applications. However, with the desire to network consumer electronic devices with PC and to play-back music recorded from the PC on CD/DVD players, these formats are also supported on wide variety of devices in the home.

2.2.3 Transport and File Formats

As discussed in previous sections, the audiovideo coding formats convert the digital media to compressed formats that are represented as binary streams. The main task of the transport and file formats is focused on multiplexing and synchronizing of these coded media streams into a single bitstream or multiple bitstreams. To perform these tasks, mechanisms are needed to distinguish the different media and also to provide timing and synchronization capabilities. In this section, we cover three such formats from the MPEG-2 and MPEG-4 standards, which hopefully will highlight different requirements and keys aspects of existing system-level formats.

The MPEG-2 Systems standard, also referred to as H.222, specifies two formats: the Transport Stream (TS) and the Program Stream (PS) [16]. Each is optimized for a different set of applications, which are discussed further below. An important feature of any system-level stream is being able to retrieve the coded media from within the transport stream, decode it, and present the decoded results. To achieve this, the system-level stream is first demultiplexed, and relevant data including both system- and media-related data would then be decoded and presented in a synchronized manner.

The TS is a system-level stream definition that is designed for communicating or storing one or more programs of coded video, audio, or other kind of data in lossy or noisy environments in which significant errors may occur. The TS is able to combine one or more programs with one or more time bases into a single stream. This is the typical system-level format used for broadcasting services around the world, including cable, satellite, and terrestrial transmission.

The program stream on the other hand is defined for the multiplexing of audio, video, and other data into a single stream for communication or storage applications. The essential difference between the program stream and transport stream is that the transport stream is designed for applications in which some level of noise must be tolerated, such as in over-the-air broadcasting, whereas the program stream is designed primarily for applications that enjoy a relatively error-free environment, such as in DVD and other digital storage applications. As such, the overhead in the program stream is less than in the transport stream.

The MP4 file format that is part of the MPEG-4 specification [17] offers a versatile container to store content in a form that is suitable for streaming. The MP4 file format can be used to store multiple variations of the same content and to select an appropriate version for delivery. The hint track mechanism supported by the MP4 file format allows supporting multiple transport formats [e.g., Real Time Transport Protocol (RTP) and MPEG-2 TS] with minimal storage overhead. Another unique characteristic of the MP4 file format is that the audiovisual media data is stored separately from its metadata, which includes timing information, the number of bytes for a video frame, file location, and so on. Also, in contrast with previous standards, the timing information is not absolute, but relative; this feature helps facilitate editing operations.

2.2.4 Profiles and Levels

Profiles and levels are critical concepts regarding media format interoperability. Profiles essentially limit the set of tools that need to be implemented from a decoder's point of view. Levels on the other hand define the complexity bounds. The profile and level combination provides a well-defined conformance point, which is needed to ensure interoperability between different implementations and to also enable testing for conformance to the standard. It should be noted that profiles and levels exist for all types of media coding formats, including video, audio, graphics, and system layers.

It is noted that profiles in the context of MPEG and other bodies that define coding standards are specified for a particular coding format. Combinations of audio and video formats are not typically defined by such standards bodies; this has traditionally been done by various industry forums. For instance, ATSC specified the use of MPEG-2 Systems, MPEG-2 Video, and AC-3 Audio for digital terrestrial broadcasting in the United States [18].

Considering the latest H.264/AVC standard, several profiles targeting different classes of applications have been defined. Each of these profiles allows a certain set of coding tools to be used for video compression. For instance, the Baseline profile only allows I- and P-slices, while the Main profile additionally allows B-slices. As another example, error-resilience tools are quite useful for applications that transmit data over noisy channels but might not necessarily be required for all applications. Therefore, such tools would be included in only a subset of profiles.

Profiles for a given standard may be defined in a hierarchical manner (e.g., as in MPEG-2 Video). In this way, decoders that comply with profiles that are higher in the hierarchy could also understand bitstreams that comply with profiles that are essentially a subset of this profile. However, a strict hierarchy is not always possible and typically depends on several factors such as the diversity of available tools, market requirements, tool complexity, and so forth. As an example, MPEG-4 Visual profiles are not defined according to a strict hierarchy.

Levels are related to the complexity and are often defined in terms of maximum bit-rates, as well as maximum frame sizes and maximum decoded picture buffer sizes, which are especially important for video decoding. Such limits are needed so that decoders could be appropriately designed and matched with target application requirements. For instance, mobile applications usually comply with levels that significantly restrict the complexity due to limited processing and display.

2.3   METADATA FORMATS

Two major classes of metadata are introduced in this section: content descriptions and usage environment descriptions. The first provides a basic understanding of the source content including the coding format, attributes, and parameters of the media. The usage environment is the space in which content is processed and consumed and characterized by several important dimensions such as terminal capabilities and network conditions. With these two pieces of information, the mismatch between source and destination is well defined.

User preferences typically play an important role in customizing and filtering of content but may also indicate a preference for a particular format. This section briefly introduces the types of descriptions that exist and discusses how they may be used to help resolve conflicts when multiple media formats are available or satisfy the destination constraints.

The final item covered in this section is electronic program guide, which is the means by which some of this metadata, the content descriptions in particular, are delivered to the home. The functionality enabled by this guide information is discussed.

2.3.1   Content Descriptions

The description of multimedia content is an extremely important piece to solve the media format interoperability problem as it provides an essential understanding of the source material to be distributed. One such standard for describing multimedia content is MPEG-7. We will discuss the most relevant MPEG-7 tools for media format interoperability here. It is recognized that alternative metadata formats do exist, especially within specialized content domains (e.g., ID3 tags for music and the Exchangeble Image File Format (EXIF) metadata format for digital photos). Because it is not practical to cover the particulars of every metadata format in this section, we aim to express the importance of particular metadata fields and what they enabled.

Among the many MPEG-7 tools available, a subset of these tools is particularly targeted toward supporting the distribution to a variety of end terminals considering other user or networking constraints. These tools are highlighted in the following, and their use in the context of home networking applications is described. In particular, we focus on the tools for describing the media format but also touch on some related tools for describing data abstractions and multiple versions of the content. Further details about the MPEG-7 standard may be found in Ref. 19.

2.3.1.1 *Media Format*   Describing the format of a media source is one of the most fundamental pieces of information that is needed to achieve interoperability. Generally speaking, the media format is specified by a variety of factors including the resolution of the image or video, the bit-rate at which the media was compressed, as well as the coding syntax that the compressed bitstream complies with (e.g., which MPEG profile/level). Most fields are common (i.e., independent of the coding syntax) and could be simply specified by integer or string values. However, the coding syntax requires special care because (1) there are many conformance points that could potentially be described, (2) the number of conformance points is growing, so the specification should be extensible, and (3) a generic mechanism to specify coding syntax of any media type, whether it be audio, video, or graphics, would be desirable.

In Part 5 of the MPEG-7 standard [20], the media coding syntax is specified using a controlled term list and classification schemes. The classification schemes defined in MPEG-7 include controlled term lists for visual, audio, and graphics formats. For the most part, the terms defined in these lists refer to specific conformance points as defined by the respective standards. For example, in the classification scheme that corresponds with visual coding formats, there exists terms to identify the various MPEG-4 Visual profiles and levels. These lists may be used as-is but may also be extended by means of a registration authority so that other coding formats, including proprietary formats or formats that have yet to be defined, could be accounted for.

2.3.1.2 *Data Abstraction*   In general, summaries provide a compact representation, or an abstraction, of the audiovisual content to enable discovery, browsing, navigation, and visualization [21]. The Summary Description Schemes (DS) in MPEG-7 enables two types of navigation modes: hierarchical and sequential. In the hierarchical mode, the information is organized into successive levels, each describing the audiovisual content at a different level of detail. The levels closer to the root of the hierarchy provide more coarse summaries, and levels further from the root provide more detailed summaries. On the other hand, the sequential summary provides a sequence of images or video frames, possibly synchronized with audio, which may compose a slide show or audiovisual skim.

The description of summaries can be used for adaptive delivery of content in a variety of cases in which limitations exist on the capabilities of a terminal or even an end-user's time. In a home networking scenario, imagine that you would like to browse the highlights from a recently recorded soccer game—either from a television set or remotely from a PDA. The summary description would essentially provide pointers into the main content to enable efficient transmission and navigation to the scenes of interest. The metadata may be generated during recording or shortly after recording has completed.

2.3.1.3 Multiple Variations

Variations provide information about different versions of audiovisual programs, such as summaries and abstracts; scaled, compressed, and low-resolution versions; and versions with different languages and modalities (e.g., audio, video, image, text, and so forth). One of the targeted functionalities of MPEG-7's Variation DS is to allow a server or proxy to select the most suitable variation of the content for delivery according to the capabilities of terminal devices, network conditions, or user preferences. The Variations DS describes the different alternative variations. The variations may refer to newly authored content or correspond with content derived from another source. A variation fidelity value gives the quality of the variation compared with the original. The variation type attribute indicates the type of variation, such as summary, abstract, extract, modality translation, language translation, color reduction, spatial reduction, rate reduction, compression, and so forth. Being able to efficiently access different variations of the content sources, which may ultimately provide a better match to the target device, has clear benefits for remote access to content. Further details on the use of variations for adaptable content delivery can be found in Ref. 22.

2.3.1.4 Transcoding Hints

Transcoding hints provide a means to specify information about the media to improve the quality and reduce the complexity for transcoding, which is the conversion process between one media format and another and elaborated on further in the next section of this chapter. Among the various transcoding hints that have been standardized by MPEG-7, we focus on the Difficulty Hint, Motion Hint, and Coding Hint. The Difficulty Hint describes the coding complexity of the original source content. This hint can be used for improved bit-rate control and bit-rate conversion; for example, from constant bit-rate (CBR) to variable bit-rate (VBR). The Motion Hint describes (a) the motion range, (b) the motion uncompensability, and (c) the motion intensity. This metadata can be used for a number of tasks including anchor frame selection, encoding mode decisions, frame-rate and bit-rate control. The Coding Hint provides high-level information about the way a video bitstream was coded, such as the distance between anchor frames and the average quantization parameter used for coding, which are not easily extracted or parsed from the bitstream itself. These transcoding hints, especially the search range hint, aim to reduce the computational complexity of the transcoding process. Further information about the use of these hints may be found in Ref. 23.

2.3.2 Usage Environment Descriptions

Content descriptions certainly provide part of the information needed to adapt content to fit a given usage or consumption environment, but we also need a description of the usage environment itself. Generally speaking, the usage environment covers a wide range of factors that might affect the optimal way that content is ultimately consumed. Undoubtedly, the most important factors are terminal capabilities and network characteristics, which are the focus of this section.

In particular, we review the metadata fields that have been specified by the MPEG-21 Digital Item Adaptation (DIA) standard [24, 25]. It should be noted that the W3C Consortium has also engaged in similar efforts to bridge this gap between multimedia content and devices. In particular, the Device Independence Working Group has completed work on the Composite Capability/Preference Profiles (CC/PP), which specifies a structure and vocabularies for device capabilities and user preferences [26]. Because the aim of this section is to emphasize the importance of metadata that describes certain attributes, such as the terminal and network, a detailed comparison of the different approaches is not given. We instead cover a select set of metadata fields and discuss their relevance to media format interoperability.

2.3.2.1 Terminal Capabilities Within the MPEG-21 DIA standard, terminal capabilities are described in terms of both receiving and transmitting capabilities. Such a description is used to satisfy consumption and processing constraints of a particular terminal. The description mainly includes codec capabilities, input–output characteristics, and other device properties, such as CPU characteristics. These various description categories are elaborated further below.

Encoding and decoding capabilities specify the format a particular terminal is capable of encoding or decoding (e.g., an MPEG profile/level). Given the variety of different content representation formats that are available today, it is not only necessary to be aware of the formats that a terminal is capable of handling, but it is sometimes important to also know the limits of specific parameters that affect the operation of the codec. In MPEG standards, the level definition often specifies such limits. However, it is possible that some devices are designed with further constraints, or that no specification of a particular limit even exists. Therefore, the codec parameters as defined by DIA provide a means to describe such limits (e.g., the maximum bit-rate that a decoder could handle).

Another important category is input–output characteristics. Display capabilities, audio output capabilities, and user interaction inputs are the key items considered under this category. Describing the capabilities of a display is obviously very important as certain limitations that impact the visual presentation of information must be taken into consideration, such as the resolution, the color capabilities, and rendering format. The same is true for audio output devices, where descriptions of frequency range, power output, signal-to-noise ratio, and the number of output channels are described. Finally, user interaction inputs define the means by which a user can interact with a terminal. With such information, an adaptation engine could modify the means by which a user would interact with resources. For instance, knowing whether a terminal has the ability to input information through a keypad or microphone may affect the interface that is presented to the user.

There are a variety of other properties that would describe a terminal, but for the purpose of this chapter, we only cover power and storage characteristics, as well as CPU benchmark measures. The power characteristics tool is intended to provide information pertaining to the consumption, battery capacity remaining, and battery time remaining. With such attributes, a sending device may adapt its transmission strategy in an effort to maximize the battery lifetime. Storage characteristics are defined by the input and output transfer rates, the size of the storage, and an indication of whether the device can be written to or not. Such attributes may influence the way that content is consumed (e.g., whether it needs to be streamed or could be stored locally). To gauge computational performance, a benchmark-based description has been specified, where the CPU performance is described as the number of integer or floating-point operations per second. With such a measure, the capability of a device to handle a certain type of media, or media encoded at a certain quality, could be inferred.

2.3.2.2 Network Characteristics Two main categories are considered in the description of network characteristics: capabilities and conditions. The capabilities define static attributes of a network, whereas the conditions describe dynamic behavior. These descriptions primarily enable multimedia adaptation for improved transmission efficiency.

Network capabilities include attributes that describe the maximum capacity of a network and the minimum guaranteed bandwidth that a network can provide. Also specified are attributes that indicate if the network can provide in-sequence packet delivery and how the network deals with erroneous packets (i.e., does it forward, correct, or discard them).

Network conditions specify attributes that describe the available bandwidth, error, and delay. The error is specified in terms of packet loss rate and bit error rate. Several types of delay are considered, including one-way and two-way packet delay, as well as delay variation. Available bandwidth includes attributes that describe the minimum, maximum, and average available bandwidth of a network. Because these conditions are dynamic, time-stamp information is also needed. Consequently, the start time and duration of all measurements pertaining to network conditions are also specified. However, the end points of these measurements are left open to the application performing the measurements.

2.3.3 User Preferences

User preferences play an important role in the way that content might be filtered or customized. In the context of media format interoperability, user preferences might suggest a preferred format for different classes of devices. MPEG-7 has standardized a collection of metadata related to user preferences, which are covered below. The basic data types have also been adopted by other standards including TV-Anytime [27] and MPEG-21 Digital Item Adaptation [25].

The UserInteraction DS defined by MPEG-7 describe preferences of users pertaining to the consumption of the content, as well as usage history. The MPEG-7 content descriptions can be matched to the preference descriptions in order to select and personalize content for more efficient and effective access, presentation, and consumption. The UserPreference DS describes preferences for different types of content and modes of browsing, including context dependency in terms of time and place. The UserPreference DS describes also the weighting of the relative importance of different preferences, the privacy characteristics of the preferences, and whether preferences are subject to update, such as by an agent that automatically learns through interaction with the user. The UsageHistory DS describes the history of actions carried out by a user of a multimedia system. The usage history descriptions can be exchanged between consumers, their agents, content providers, and devices, and may in turn be used to determine the user's preferences with regard to content.

2.3.4 Electronic Program Guide

The Electronic Program Guide (EPG) is essentially a description of programming information. In its most basic form, live television programs are described by program title, channel, and time. The display format that most people are familiar with today is a simple grid-like structure that could be used to browse the programs at a given time on a given channel. When a program is selected, the device could tune to the channel or schedule a recording of the selected program.

The EPG data that is available today is significantly richer and could include a list of actors in a program, release and production information, the genre, parental guidance and rating data, as well as metadata about the media formats. Links to related material such as trailers and reviews may also be included along with unique identifiers and associated rights information. With this type of information available, new applications could be developed to provide better access to the increasing amounts of content and improve the overall user experience.

There exist several standards that specify metadata to support EPG services. One has been developed by the TV-Anytime Forum and has been published as a European Telecommunications Standards Institute (ETSI) standard [27]. Another specification is being developed by the Consumer Electronics Association (CEA) and is built upon the TV-Anytime specification [28]. The main target for the CEA specification is the U.S. market, so the different requirements on describing the distribution networks and content services must also be accounted for. It should be noted that both specifications reference data types defined by MPEG-7, including those that describe the media format. Whether such metadata is being delivered as part of an integrated content service or provided as a separate service to users, its existence in the home will play an important role in achieving media format interoperability.

2.4 MEDIA ADAPTATION

The general process of converting compressed streams in a given format to another target format is referred to as media adaptation or transcoding. The term *format* covers various aspects of the compressed media including spatial and temporal resolutions for video, sampling rates for audio, bit-rates, bitstream syntax, and so on. The media format usually complies with a certain profile and level as discussed earlier in Section 2.2.4.

The aim of this adaptation or transcoding process is to ultimately enable transfer of the original bitstreams over the appropriate network or interface and play back on a target device. Most of the transcoding research has focused on two key aspects: (1) keeping the complexity as low as possible and (2) minimizing signal degradation. The reference transcoding method is typically referred to as the cascaded approach. This cascaded approach simply decodes the original signal, performs any intermediate processing, and re-encodes to the target format. This method has certain advantages. For instance, it is able to minimize losses in quality. Also, it uses conventional and unmodified decode and encode operations. The main disadvantage of the cascaded approach is that it is computationally intensive and costly, especially in the case of video conversion. Before discussing alternative transcoding approaches, some example scenarios are discussed.

Figure 2.2 shows several typical transcoding operations. In this example, the original video is encoded in an MPEG-2 format (Main Profile at Main Level = MP@ML) at 5.3 Mbit/s. The input resolution is $720 \times 480i$ (interlaced) and the temporal rate is 30 frames per second (fps); the original audio is encoded in an AC-3 format at 192 kbit/s. Both audio and video files are multiplexed into an MPEG-2 Transport Stream (TS). This package represents a typical standard-definition broadcast stream. In one instance, the original video is transcoded to an H.264/AVC format that complies with the Baseline Profile (BP) at level 1.3. The bit-rate of the transcoded video is 769 kbit/s; the frame rate has been reduced down to 10 fps

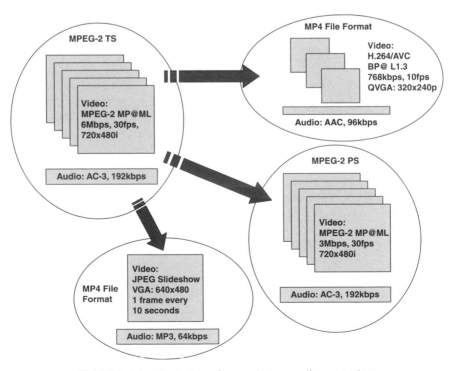

FIGURE 2.2   Illustration of example transcoding operations.

and the spatial resolution down to Quarter Video Graphics Array (QVGA) resolution or 320×240. The audio has also been transcoded to AAC format at a bit-rate of 96 kbit/s. These two media files are wrapped in an MP4 file format, which is suitable for streaming and is supported on a variety of mobile devices. In a second instance, the original broadcast stream is transcoded for long-term storage. In this case, only the bit-rate of the video is reduced to 3 Mbit/s to save on disk space. Also, rather than storing the media as an MPEG-2 TS, the system layer is converted to an MPEG-2 Program Stream (PS) format. In a final instance, the original stream is transcoded to a slide show of JPEG images, where each image is VGA resolution and refreshed every 10 seconds. The audio has also be transcoded to a MP3 at a rate of 64 kbit/s. Both audio and video are multiplexed and synchronized as an MP4 file.

As mentioned above, transcoding techniques are aimed at avoiding the full decoding and re-encoding of streams to satisfy network conditions and terminal capabilities. Although the cascaded approach is feasible for audio due to the lower data rates and amount of computation required to decode and re-encode, the complexity to fully decode and re-encode video has motivated researchers to find alternative low-complexity transcoding solutions that maintain high picture quality. One general approach has been to reuse existing information from the input stream in the output stream. For example, the motion vectors from the input video bitstream could be reused in the output. If the spatial resolution is being reduced, it is possible to map the motion vectors to the output scale and perform a simple refinement operation. Another approach has been to simplify the cascaded decode and encode architecture. Mathematically equivalent architectures that eliminate certain processing components have been proposed and are able to achieve transcoded picture quality that is very close to the reference method in most cases. Extensive reviews of video transcoding technology do exist [27–30], and readers are referred to these articles for further information.

2.5 MANDATORY MEDIA FORMAT PROFILES

The Digital Living Network Alliance (DLNA) is an organization that has been formed with the goal to establish a platform of interoperability based on open and established industry standards. The alliance has released design guidelines to achieve interoperability among home devices, which among other things deals with media format interoperability [31]. The approach that has been taken is to define a set of mandatory media format profiles that all devices within a device class or category are required to support. In this context, a profile defines the combination of audiovisual (AV) compression formats, media-specific attributes and parameters, as well as system-level format and any other information that would sufficiently describe the encoded content. The intention of such a model is to achieve a baseline for home network interoperability. Optional media formats are also specified to allow for broader support of other popular media formats.

The latest volume of Media Format Profiles as defined by DLNA specifies the detailed guidelines to enable interoperability between DLNA devices in the digital home [32]. An example profile would define the AV media formats as well as the encapsulation or system-layer format. For instance, a profile ID of `AVC_MP4_BL_L2_CIF30_AAC` indicates that the video coding format is compliant to H.264/AVC Baseline Profile at Level 2. The picture resolution is Common Intermediate Format (CIF) (352×288) and the maximum frame rate is 30 Hz. The audio format is AAC with a maximum bit-rate specified by DLNA as 128 kbit/s. The encapsulation for this DLNA profile is designated as MP4. As one could imagine, given all the possible AV media and encapsulation formats, the number of profiles is quite large. The sets of mandatory and optional formats for a given device class are defined using such profiles.

In order to support interoperability between devices of different classes, it is expected that some translation between the required media format profiles of different device classes would be needed. Additionally, DLNA specifies rules about conversion between optional and mandatory formats to ensure that content can be enjoyed on all compliant devices.

2.6   MEDIA FORMAT INTEROPERABILITY: AN EXAMPLE

In this section, we provide a simple example that attempts to pull together some of the key concepts that have been covered in this chapter. Figure 2.3 illustrates how media

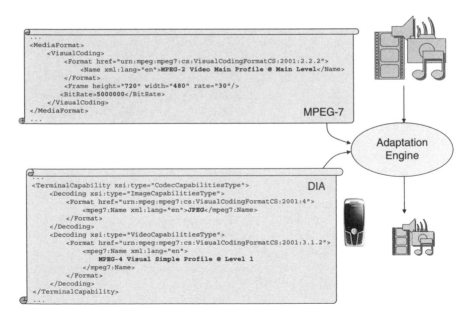

FIGURE 2.3   Achieving media format interoperability, where adaptation is based on both MPEG-7 metadata describing the content and MPEG-21 DIA metadata describing the terminal capabilities.

format interoperability is achieved through metadata about the content and target device and a media adaptation engine.

First, consider a high-quality full frame-rate video with resolution of 720×480 pixels that is encoded according to the MPEG-2 Main Profile @ Main Level format at a bit-rate of 5 Mbit/s. The source of this content may be broadcast or DVD. Regardless of the content source, an MPEG-7 description corresponding with this content would state the properties of this media. If the content originated from a broadcasting service, this metadata might be carried as part of the EPG. If it is not delivered as part of the broadcast service, it may also be generated locally within a home network device by parsing the sequence-level headers of the video bitstream.

Next, assume that a mobile terminal wants to access this video but is only capable of decoding JPEG images and video encoded in the MPEG-4 Visual Simple Profile @ Level 1 format. Furthermore, it only has a display resolution of 176×144 luminance pixels. The MPEG-21 DIA description of this terminal would state properties of the terminal itself. Such a description may originate from the device itself or a reference to the description could be provided (e.g., by means of a URL).

In order for the mobile terminal to render this high-quality video material on its screen, adaptation from the original MPEG-2 Video format to the destination MPEG-4 Visual format must be performed. As discussed earlier, the Simple Profile of MPEG-4 Visual indicates the syntax that the adapted bitstream must conform to, and the Level indicates defined limits on various aspects of the stream, such as the maximum number of macroblocks per second (MB/s) and the maximum bit-rate. In the case of Level 1, the maximum number of MB/s is 1485, which corresponds with a typical spatial resolution of 176×144 pixels and temporal resolution of 15 Hz, and the maximum bit-rate is 64 kbit/s. It should be clear that the output of the adaptation engine must conform to these specifications in order for the content to be consumed on the mobile terminal.

As shown by this simple example, the MPEG-7/21 metadata provides the adaptation engine with necessary information required to convert the media to the target format. In doing so, media format interoperability is achieved. As the heterogeneity of devices and the number of source coding formats increases, being able to match the content to the terminal becomes quite a significant task, and the use of metadata to guide the adaptation engine becomes essential.

2.7 CONCLUSIONS

This chapter discussed the challenge of achieving media format interoperability among home network devices. Some common AV media formats and system-layer formats were reviewed, and several technical solutions have been presented that aim to bridge the gap between multimedia content from a variety of sources to the wide range of receivers with different capabilities. It is clear that the diversity of receiver terminals, networks, and content formats necessitates standardization of not only the media formats themselves but also the metadata formats that describe these formats and the usage environment in which these formats are processed and consumed. The importance of transcoding between different media formats has been highlighted, and some of the key technologies for computationally efficient transcoding of video have been covered. Even with standards that attempt to define a baseline set of media profiles for different classes of devices, conversion of media formats will still be necessary. The hope is that all of these layers of interoperability and mechanisms that help to facilitate interoperability will be transparent to consumers. Consumers should soon be able to seamlessly enjoy both personal and commercial content on any authorized and compliant device within their network.

REFERENCES

1. ISO/IEC 10918-1:1994, Information technology—Digital compression and coding of continuous-tone still images: Requirements and guidelines.

2. ISO/IEC 15444-1:2004, Information technology—JPEG 2000 image coding system: Core coding system, 2nd ed.

3. ITU-T Recommendation H.261, Video Codec for Audiovisual Services at px64 Kbit/s, March 1993.

4. ISO/IEC 11172-2:1993, Information technology—Coding of moving pictures and associated audio for digital storage media at up to about 1.5 Mbit/s—Part 2: Video.

5. ISO/IEC 13818-3:1998, Information technology—Generic coding of moving pictures and associated audio information—Part 3: Audio, 2nd ed.

6. ITU-T Recommendation H.263, Video Coding for Low Bit Rate Communication, May 1996.

7. ISO/IEC 14496-2:2004, Information technology—Coding of audio-visual objects—Part 2: Visual, 3rd ed.

8. ISO/IEC 14496-10, Information technology—Coding of audio-visual objects—Part 3: Advanced Video Coding (AVC), 3rd ed.

9. SMPTE 421M-2006, Television—VC-1 Compressed Video Bitstream Format and Decoding Process.

10. AES3-2003, AES Recommended practice for digital audio engineering—Serial transmission format for two-channel linearly represented digital audio data (Revision of AES3-1985 and AES3-1992, including subsequent amendments).

11. ISO/IEC 11172-3:1993, Information technology—Coding of moving pictures and associated audio for digital storage media at up to about 1.5 Mbit/s—Part 3: Audio.

12. ISO/IEC 13818-7:2006, Information technology—Generic coding of moving pictures and associated audio information—Part 7: Advanced Audio Coding (AAC), 4th ed.

13. ISO/IEC 14496-2:2005, Information technology—Coding of audio-visual objects—Part 3: Audio, 3rd ed.

14. T. Ebrahimi and F. Pereira, Eds., *The MPEG-4 Book*. Prentice Hall, Englewood Cliffs, NJ, 2002.

15. ATSC Standard A/52A, Digital Audio Compression (AC-3) Rev A, Advanced Television Systems Committee, August 2001.

16. ISO/IEC 13818-1:2000, Information Technology—Generic coding of moving pictures and associated audio information—Part 1: Systems, 2nd ed.

17. ISO/IEC 14496-14:2003, Information Technology—Coding of Audio-Visual Objects—Part 14: MP4 file format.

18. M.S. Richer, G. Reitmeier, T. Gurley, G.A. Jones, J. Whitaker, and R. Rast, "The ATSC Digital Television System." *Proc. IEEE*, Vol. 94, no. 1, pp. 37–43, January 2006.

19. B.S. Manjunath, P. Salembier, and T. Sikora, Eds., *Introduction to MPEG 7: Multimedia Content Description Language*. John Wiley & Sons, Hoboken, NJ, 2002.

20. ISO/IEC 15938-5, Information technology—Multimedia Content Description Interface—Part 5: Multimedia Description Schemes.

21. P. van Beek, J.R. Smith, T. Ebrahimi, T. Suzuki, and J. Askelof, "Metadata-driven multimedia access." *IEEE Signal Processing Magazine*, Vol. 20, no. 2, pp. 40–52, March 2003.

22. R. Mohan, J.R. Smith, and C.S. Li, "Adapting multimedia internet content for universal access." *IEEE Trans. Multimedia*, Vol. 1, no. 1, 104–114, March 1999.

23. P. Kuhn, T. Suzuki, and A. Vetro, "MPEG-7 transcoding hints for reduced complexity and improved quality." *Proc. Int'l. Packet Video Workshop*, Kyongju, Korea, April 2001.

24. A. Vetro and C. Timmerer, "Digital item adaptation: Overview of standardization and research activities." *IEEE Trans. Multimedia*, Vol. 7, no. 3, pp. 418–426, June 2005.

25. ISO/IEC 21000-7:2004, Information Technology—Multimedia Framework—Part 7: Digital Item Adaptation.

26. W3C Recommendation, "Composite Capability/Preference Profiles (CC/PP): Structure and Vocabularies 1.0," January 2004.

27. ETSI TS 102 822-3-1, Broadcast and On-line Services: Search, select, and rightful use of content on personal storage systems ("TV-Anytime"); Part 3: Metadata; Sub-part 1: Phase 1—Metadata schemas, version 1.3.1, January 2006.

28. Draft CEA-2033, OpenEPG—A specification for Electronic Program Guide Data Interchange, 2006.

29. A. Vetro, C. Christopoulos, and H. Sun, "An overview of video transcoding architectures and techniques." *IEEE Signal Processing Magazine*, Vol. 20, no. 2, pp. 18–29, March 2003.

30. J. Xin, C.W. Lin, and M.T. Sun, "Digital Video Transcoding." *Proc. IEEE, Special Issue on Advances in Video Coding and Delivery*, December 2004.

31. Digital Living Network Alliance Home Networked Device Interoperability Guidelines Version: v1.5, Volume 1, Architecture and Protocols, 2006.

32. Digital Living Network Alliance Home Networked Device Interoperability Guidelines Version: v1.5, Volume 2, Media Format Profiles, 2006.

3

MEDIA DESCRIPTION AND DISTRIBUTION IN CONTENT HOME NETWORKS

EDWIN A. HEREDIA

Fueled by the continuous price reduction of wired and wireless network infrastructure components, networking technologies based on Ethernet and 802.11 (a/b/g) are moving rapidly from the corporate world into the home environment. Unlike the corporate world, media and entertainment become the most attractive applications in the home; applications that due to their special characteristics impose additional features and requirements on the operational functionality of the network. Currently, home networks are used principally for sharing Internet access or for sharing printers. It is expected, however, that in the near future the most important applications for home networking will be related to content distribution.

Home networks will capture content from multiple sources like optical media, digital television, cellular phone networks, personal video cameras, the Internet, and several others. Content items coming from these heterogeneous sources will have to be distributed throughout the home for final consumption (Fig. 3.1). Within the home, the devices that will render this content also show heterogeneous decoding capabilities. For example, the hardware in a digital television (DTV) receiver is fitted to decode television video but could have problems decoding Internet-based video streams. Other devices like PCs, cell phones, PDAs, and personal music players also have this focused scope in terms of decoding capabilities.

Having heterogeneous content items for distribution to a collection of receivers with heterogeneous decoding capabilities creates a problem; especially considering

Technologies for Home Networking. Edited by Sudhir Dixit and Ramjee Prasad
Copyright © 2008 John Wiley & Sons, Inc.

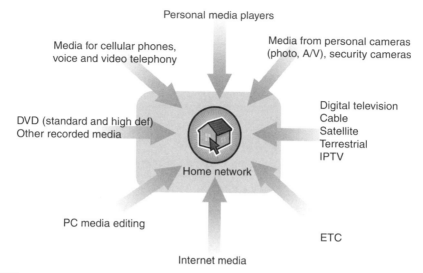

FIGURE 3.1 The content home network as the convergence center for multiple hetero-geneous media sources.

that the number of coding formats, file formats, copy protection formats, and streaming formats continues to increase with time. The average home user expects his or her entertainment devices to work with high quality and with little user effort. A typical home user will spend little or no effort trying to understand differences in codec or packet formats between the different sources. For this reason, solutions are required to cope with the multiplicity of formats preferably in such a way that the process becomes completely transparent to the user. Software agents specialized in translation or conversion between formats, which perform their tasks transparently with respect to the user, could become part of the solution.

In this chapter, we discuss the problem of diversity in media formats in content home networks and introduce a mark-up construct to be used for the description of Content Format Variants and for the recognition of device processing capabilities during server–client negotiations in content home networks. Section 3.1 attempts to explain the rationale for the diversification of media formats into multiple format variants. Section 3.2 describes a content home networking architecture and introduces the main components or capabilities of networked media devices. Next, in Section 3.3, we review the problem of coping with multiple format variants in the home. In Section 3.4, we explore the requirements for describing format variants for the representation of media objects and for the declaration of device processing capabilities. Section 3.5 presents a description language that could be used to satisfy the requirements. An early version of the ideas presented in this chapter appears in Ref. 1.

3.1 DIVERSIFICATION OF MEDIA FORMAT VARIANTS

Although it is probably difficult to pinpoint precisely when we started migrating from analog to digital media, it is nevertheless true that this migration has accelerated substantially during the past 10 years: from vinyl disks to CDs, VHS tapes to DVDs, cassette tapes to flash players. Despite this tremendous progress, there are still many sources of analog signals. In most parts of the world, people listen to analog music in their AM or FM receivers and watch analog television in their National Television System Committee (NTSC), Phase Alternating Line (PAL), or Séquentiel Couleur à Mémoire (SECAM) receivers. However, like the other technologies mentioned before, these mass-market media technologies are already migrating into digital formats, and it is expected that within the next decade, in the industrialized countries, radio and television should have completed its migration process.

At the same time as the digital media market grows, network infrastructure devices have become cheaper and more accessible to the average user. The prices of wireless and/or wired network infrastructure devices like hubs, switches, routers, and access points have been going down fueled in part by strong and quick corporate adoption. This price reduction has made it possible for a large number of homes to deploy initial home networks whose main purpose is typically to share PCs, printers, or access to the Internet. In other words, the main purpose of existing home networks seems to be similar to the enterprise environment: share productivity resources among many potential users. This paradigm, though, seems to be on the verge of changing.

The convergence of digital media sources and cheap networking components in the home creates an opportunity for the deployment of media-related or content home networks. Unlike the enterprise model, the primary objective of such networks would be the exchange of digital media; in the form of audio, video, images, and combinations of media objects (multimedia). After all, it is clear that people dedicate and invest much more time to entertainment rather than office matters when spending time at home.

However, content home networks bring new problems and new challenges in terms of technology and usability requirements with respect to the more conventional enterprise-oriented networks. For example, consider the case of video delivery over conventional networks and video delivery over cable. Home users are thoroughly familiar with the latter, and when exposed to networked video, they subjectively expect the same level of quality and performance. Video glitches due to network congestion or long periods of fuzzy images due to increased compression (performed for example as a means to minimize video bandwidth during congestion) would be examples of quality/performance problems that, unlike office users, home users would not tolerate. There is an opinion that because users tolerate the lower quality of Internet streaming video, they will also tolerate the same type of artifacts when watching video streamed over a home network. However, we believe that this is not the case. Users purchase high-quality content in the form of optical media (DVDs) or through subscriptions (e.g., cable TV), and they expect to see a high-quality output regardless of whether the content is played locally or over a home network. Users will show little tolerance for home networking technologies that degrade the high-quality content that they purchase. In summary, a properly designed content home network has to deliver video (or audio or any form of multimedia content) with a quality that at least matches existing home user experience with home digital media.

One of the existing user expectations relates to the ability to play media objects everywhere independent of the type of player device. A cassette tape recorded in Europe would play in any U.S. tape player. This expectation becomes difficult to maintain in the digital world. Even in the analog world, this expectation has been broken historically a number of times; for example, when video tapes and video players had to be identified as belonging to the VHS or Betamax formats. Even after the prevalence of VHS over Betamax, this expectation could not materialize due to differences in analog television formats created by standards like NTSC, PAL, and SECAM. A user that recorded a television program in Europe (PAL) would have problems playing this content in U.S. players (NTSC). Players designed to decode multiple formats eventually alleviated this problem.

In the digital world, this problem becomes much more serious because it is extremely simple to change, improve, constrain, and extend digital media standards creating multiple variants of a given format. Consider for example the case of high-quality digital video. The original standard created for this purpose was MPEG-2 (Moving Picture Experts Group) [22–4]. MPEG-2 already used the concept of profiles and levels as the means to allow applications to choose one or a few resolutions. A digital television application, for example, would choose profiles and levels that matched its high-definition requirements, and a video phone application could choose other profiles and levels that matched the smaller resolutions that video phones were supposed to have in those days.

In addition to profile and levels, the original MPEG-2 specifications already introduced two different methods to encapsulate the encoded signals: Program Streams and Transport Streams [2]. The Transport Stream layer defined the packetization of MPEG-2 encoded signals using short-length packets (188 bytes) that were considered better suited for error-prone communication links such as television transmissions. Program Stream packets (or packs, as they are called in the specifications), on the other hand, were designed to have a much larger size, to match what in the early MPEG-2 days was considered a good solution for optical media. In consequence, from the very beginning the original MPEG-2 specification introduced the notion of heterogeneous format variants defined to match specific application requirements. One application could choose a subset of profile and levels using Program Streams (one variant). Another application could choose a different subset of profile and levels using Transport Streams (another variant). These two applications would not normally interoperate. Figure 3.2 shows a diagram of the diversification of the MPEG-2 standard into multiple variants due to new standards that adopt the original standard and define constraints and extensions.

Definition: A *Content Format Variant* is defined here as a change in the set of encoding format parameters, a change in the selected audio or video companion components, or a change in encapsulation and/or multiplexing formats. A Content Format Variant is generated typically with the purpose of adapting

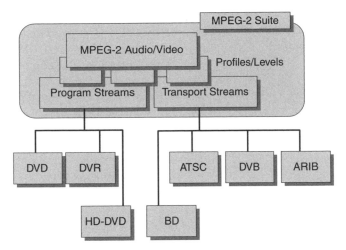

FIGURE 3.2   The adoption and diversification of MPEG-2 into multiple Content Format Variants.

baseline codec specifications to satisfy the requirements of a different application or a different usage scenario.

Using the definition introduced here, it is possible to say that the original MPEG-2 specification already presented the possibility of multiple Content Format Variants. This is by no means a flaw in the design, and instead it is seen as the natural way to cover more usage scenarios using the same standard. In fact, most if not all of existing media standards and technologies follow the same approach. For example, Microsoft's Windows Media Video technology, which has been recently standardized as the VC-1 (Video Codec 1) [5], defines profiles and levels to be used by small-screen devices like cell phones, but it also defines profiles and levels for high-definition video applications.

However, the MPEG-2 story does not end with Program Streams and Transport Streams. Due to its successful adoption by many industry segments, new Content Format Variants were introduced when standards for DVD and digital television were introduced. These standards further constrain or extend the original variant to add components required for the application. In the case of digital television, variations due to geographic interests have led to at least three standards: Advanced Television Systems Committee (ATSC) developed in North America, Digital Video Broadcasting (DVB) developed in Europe, and Association of Radio Industries and Businesses (ARIB) developed in Japan. These digital television standards in turn create new variants due to multiple usage scenarios. For example, the DVB specification is used in some cases with MPEG Layer 2 audio (e.g., in several of Europe's terrestrial DTV implementations), but in other cases it is used with AC-3 (Arc Consistency (algorithm) 3) audio (e.g., several of Australia's terrestrial high-definition DTV implementations).

Similarly, the MPEG-2 variant used in conventional DVDs has introduced particular constraints and extensions satisfying the needs of this specific successful application. Current attempts at extending the DVD formats to incorporate high-definition video (e.g., the work toward HD-DVD and BluRay Disc) will undoubtedly create new variants that will need to be taken into account within the home network. Figure 3.2 illustrates the concepts described here showing the progressive diversification of MPEG-2 and the generation of multiple Content Format Variants.

A similar evolution occurs for all the other audio and video formats either standardized or proprietary. An initial attempt by an industry consortium to classify the number of format variants that could exist in a content home network has resulted in about 260 variants that may have some relevance in the home [6, 7]. This number counts the variants resulting from image coding (Joint Photographic Experts Group (JPEG) and Portable Network Graphics (PNG)), Audio Coding (AC-3, Adaptive Multi-Rate (AMR) (audio compression), Adaptive Transform Acoustic Coding (ATRAC3plus), Linear Pulse Code Modulation (LPCM), MPEG-1 audio layer 3 (MP3), MPEG-4, and Windows Media Audio (WMA)), and Audiovisual (content) (A/V) coding (MPEG-1, MPEG-2, MPEG-4, and WMV). In fact, even this number seems small. It is expected that this number will increase significantly after adding variants derived from high-definition DVD applications (HD-DVD and BD), cell phone video services (broadcasting and downloading), Internet Protocol Television (IPTV), protected distribution, Internet-based formats, and others. Section 3.4 presents a more detailed analysis of the variants that could exist in a home network.

3.2 CONTENT HOME NETWORK ARCHITECTURE COMPONENTS

From the perspective of media distribution, a content home network is a collection of network devices that process media content binaries to provide an entertainment experience to users. A home network always has media sources and media sinks. Each physical device in a home network can have multiple components that provide certain capabilities in the network. Examples of components include media servers, media receivers, media storage, and so forth. In this section, we attempt to classify the different components that could be observed in a content home network.

A device that includes a media server (MSERV) component makes content available to any other devices in a home network. A device that includes a media receiver (MRCVR) component takes content from the network for some purpose that could include rendering. A device that includes a media storage (MSTO) component is capable of keeping a binary copy of some content item for longer periods of time.

Media capture (MCAP) components take content from the outside environment (broadcast, Internet, optical media, etc.) with the purpose of making such content available in the home network. Naturally, they need an MSERV to fulfill this purpose. Examples of devices that include an MCAP component are

- A DTV receiver that captures broadcast signals and makes digital video signals available within the content home network.

- A digital camera that captures images outside the home and bring those images to the content home network (an example of devices with intermittent connections to the home network).
- A home gateway that brings Internet or IPTV content to the home.

A control point (CP) component included in the functionality of some device would be able to select devices in the network for communications and control. Additionally, a CP component would be able to browse and select content from the network.

In addition, other components exist in the form of media converters (MCONV). A device that implements an MCONV will be able to receive content in some Content Format Variant and deliver the same item in a different variant. The conversion process could happen in real time or off-line.

Another component in the network is the media encoder (MENC). A device that implements an MENC is capable of taking raw data as an input and generating binary streams encoded with a particular Content Format Variant at the output. Finally, a device that implements a media decoder (MDEC) will be able to get encoded content and convert it into raw data possibly (and most likely) for the purpose of rendering the content.

For example, consider the case of a networked digital video recording device that takes analog television signals for storage (time-shifted viewing) and, in this case, for exposing those programs to the content home network. This device includes an MCAP component because it is used to bring content from the outside to the inside of the home network. This device also includes an MENC because it takes analog TV signals as an input and outputs digital video signals. This device also includes an MSTO component because it is capable of storing digital video signals for later viewing. This device also includes an MSERV if it is capable of exposing and transferring digital video signals to other devices in the network. This device could include a CP component to select potential devices in the network capable of acting as receivers or converters. This device includes an MDEC component because it is capable of decoding the stored video signals and displaying video on a screen.

Table 3.1 summarizes the roles of the different components that participate in media distribution within the content home network. As another example of the content home network components defined in this section, Figure 3.3 illustrates one case of a home network where these components interact to provide an entertainment experience. Figure 3.3 shows four devices, each of which implements multiple components. The set-top-box device, for example, implements a capture (MCAP) component and a server (MSERV) component. The capture component will be used to import digital television signals (from terrestrial, cable, or satellite sources). The server component will be used to make these signals available to all other devices in the network. Notice that this device does not include an MENC probably because it is a digital-only device that takes MPEG-2 signals and makes those available to the home.

The first PC in Figure 3.3 captures content from the Internet (thus, it implements MCAP), stores content on its hard disk (thus, it also implements MSTO), and is

TABLE 3.1   Summary of Content Home Network Components and Their Roles

Component	Name	Component Role
MSTO	Media storage	Stores content for possible later consumption within a home network
MCAP	Media capture	Captures content from external sources with the intent of making it available internally within the home network
MSERV	Media server	Transfers content to networked media receivers (MRCVRs)
MRCVR	Media receiver	Receives content from networked media servers (MSERVs) for any of multiple purposes like storing, rendering, etc.
CP	Control point	Performs device selection, device control, and orchestrates the transfer of content from MSERVs to MRCVRs
MCONV	Media converter	Converts content from one Content Format Variant to another
MENC	Media encoder	Converts raw content data into encoded streams that use certain Content Format Variant
MDEC	Media decoder	Converts encoded media content (using some Content Format Variant) into raw data possibly for consumption

capable of making content available to the network (MSERV). The DTV set in the figure implements an MRCVR component to receive signals from the network. It also implements an MDEC to convert encoded signals into raw data for rendering. In this example, the DTV set also implements a CP to select content from servers, select network conversion operations, etc.

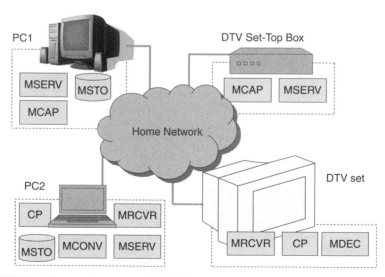

FIGURE 3.3   Content home networked devices showing examples of components that they may implement.

Finally, the laptop PC in Figure 3.3 implements a converter component (MCONV). It implements an MRCVR to receive content from the network and an MSERV component to make the converted content available to other devices in the network. It implements its own CP component probably with the purpose of selecting content from servers for conversion and storage (MSTO). Notice that the PC does not implement an MDEC, which implies that the laptop in this case is not used for decoding and rendering.

We envision a future where each of the components implemented by a certain device could become available to the network. In other words, any device in the network should be able to use any of the components that exist in other devices as long as they are idle. For example, a PC that needs to decode MPEG-2 encoded content into raw data could actually solicit this service from an idle DTV set. Similarly, a DTV device could solicit the services from a PC in order to convert content from some nonfamiliar variant into a variant that the DTV device can decode.

In this scenario, some networked devices will be able to advertise themselves as providers of content-processing services. In doing so, the devices themselves will help solving the variant multiplicity problem. Before this scenario could be realized in practice, some standardized protocols need to be developed as a baseline. Protocols like (1) advertisement of content-processing services, (2) service queries, (3) service requests and responses, and (4) data exchange protocols. In this chapter, we do not attempt to cover all the protocols that would be required to enable content-processing services, but we concentrate on the ability to describe Content Format Variants as explained in subsequent sections of this chapter.

Given that content home networked devices will typically come from multiple manufacturers using multiple types of hardware and software components, the ideal place to develop the whole suite of protocols to enable this scenario would be a standards organization dedicated to this matter. Notice that the protocols mentioned above provide only a minimal list of tasks. In order to transfer commercial content, additional topics that would require solution include, of course, the establishment of secure sessions, the transference of protected content, and others.

3.3 CONTENT FORMAT VARIANTS IN THE HOME

A content home network can be seen as the convergence center where multiple lines or sources of digital technologies come together. An example of a digital technology line is the DVD disk. For years, users have enjoyed the ability to watch MPEG-2 encoded digital movies in their DVD players. Another example of a digital technology line would be Internet radio. Again, for a number of years already, users have been using a PC with a browser to access radio stations that transmit, for example, Windows Media Video (WMA) encoded programs over the Internet. Many other similar digital technology lines are already found in devices that could belong to a home network.

This means that the home network behaves as the convergent hub where all these digital technology lines come together for the benefit of users (Fig. 3.1). However,

TABLE 3.2   Summary of Image Format Variants Derived from Refs. 6 and 7

Image Family	Feature Variations	File Format Variations	Total Number of Variants
JPEG	Small (up to 640 × 480), medium (up to 1024 × 768), large (up to 4096 × 4096), icons, thumbnails	EXIF (1.X or later), JFIF 1.02	6
PNG	Large (up to 4096 × 4096), icons, thumbnails	PNG file format	4

this convergence of a potentially large number of digital technology lines creates a problem because each line brings multiple Content Format Variants.

We already mentioned in Section 3.1 that an initial count and compilation of Content Format Variants relevant for home networks shows a total of about 260 variants [6, 7]. It is expected that this number will grow substantially with time when additional support is included for new standards, new usage scenarios, new implementations, and so forth. For example, the work in Refs. 6 and 7 does not include yet the combinations resulting from having multiple ways of encrypting content and the multiple ways of carrying rights-management licenses for distribution of commercial content. Tables 3.2, 3.3, and 3.4 summarize the Content Format Variants derived from the compilation work carried out in Refs. 6 and 7 for image, audio, and A/V, respectively.

Coping with this diversity of Content Format Variants is one of the most interesting problems in home networking especially because one of the goals of home networking has always been to provide a "plug and play" experience for networked devices. In other words, we would like users to connect their devices and be able to browse and experience content without the need to know about encoding features and device decoding capabilities.

TABLE 3.3   Summary of Audio Format Variants Derived from Refs. 6 and 7

Audio Family	Feature Variations	File Format Variations	Total Number of Variants
LPCM	Mono, stereo	Raw data	1
AC-3	Stereo, up to 5-channels	Raw data	1
MPEG4	Mono, stereo, 5-channels, 7-channels, AAC-LC, AAC-LTP, HE-AAC, BSAC	MP4, 3GPP, ISO, ADTS	19
AMR	Baseline, WB+, mono, stereo	MP4, 3GPP	2
ATRAC3plus	Mono, stereo, multichannel	ATRAC3plus	1
MP3	Mono, stereo, extensions for low sampling rates and bit-rates	MP3 file format	2
WMA	WMA and WMA Pro (up to 7-channels)	ASF	3

TABLE 3.4 Summary of A/V Format Variants Derived from Refs. 6 and 7

A/V Family	Video Variations	Audio Variations	File Format Variations	Total Number of Variants
MPEG1	CIF resolutions, CBR	MPEG1	MPEG1 System	1
MPEG2	DVD PAL-compatible, DVD NTSC-compatible, ATSC-compatible, DVB-compatible, Standard Def, High Def.	AC-3, Extended AC-3, LPCM, MPEG-1 (Layers 1 and 2)	DVD-VR Program Streams, ATSC Transport Streams, DVB Transport Streams. Original TS or with an additional zero or non-zero time stamp per TS packet. Elementary streams for RTP.	38
MPEG4 Part 2	SP at levels L0, L0b, L1, L2, L3. ASP at levels L0, L1, L2, L3, L3b, L4, L5. CO at levels L1, L2. H.263 with Profile 0 Level 10.	AAC-LC, HE-AAC, AAC-LTP, AMR, AMR WB+, MP3, G.726, MPEG-2 Layer 2, ATRAC3plus, AC-3, mono, stereo, 5-channels, 7-channels	MP4, ASF, MPEG-2, Original Transport Streams or with an additional zero or non-zero time stamp per TS packet. Elementary streams for RTP	54
MPEG4 Part 10 H.264	MP at levels L1, L1.1, L1.2, L.13, L2, L2.1, L2.2, L3, L4. BP at levels L1, L1b, L1.1, L1.2, L2.	AAC-LC, HE-AAC, AAC-LTP, AMR, AMR WB+, MP3, MPEG-2 Layer 2, ATRAC3plus, AC-3, Extended AC-3, mono, stereo, 5-channels, 7-channels	MP4, 3GPP, MPEG-2, Original Transport Streams or with an additional zero or non-zero time stamp per TS packet. Elementary streams for RTP	119
WMV	WMV9 Main Profile at levels: Medium, High. Simple Profile at levels: Low, Medium	WMA and WMA Pro (up to 7 channels), MP3	ASF, elementary streams for RTP.	9

Several strategies could be used to solve or at least mitigate the impact of the multiplicity of format combinations in a home network. One way of solving this problem would be to define a single encoding format that all home networked devices could later use and understand. This solution assumes that content at the periphery of the home network will be converted into a single Content Format Variant for distribution within the network. A major problem of this approach is that typically one solution does not fit well with all needs. The optimum variant actually depends on several things like type of rendering device, type of network connection, and so forth. Expensive A/V devices for example would prefer to receive high-quality, high-definition A/V in its original variant without any re-encoding. Devices connected using wireless 802.11b would probably prefer to receive content re-encoded using the most bandwidth-efficient compression variants.

A separate trend that has been observed in practice, especially for software-based media players like Microsoft's Windows Media Player or Apple's QuickTime, consists in implementing support for a large number of variants. Support for newer codec variations could be added over time through usual software upgrades. Users then employ these tools to play content transparently, without knowing about formats and/or player features. Although this solution works well for computers, it becomes more problematic for dedicated consumer electronics devices that might not have the flexibility for continuous updates. Recent standards like HD-DVD or BluRay Disc introduce requirements to support multiple codecs. As a result, Application Specific Integrated Circuit (ASICs) that support multiple Format Variants are becoming available; thus, consumer electronics devices in the near future should also be capable of supporting multiple variants.

At the end, the most likely scenario will be one where devices in the home will be capable of decoding one or more variants but not all of the possible codec format combinations. At the same time, some devices in the home will be able to transform content from certain variants to others. In a fully cooperative network, when a player device is interested in playing content that comes in an unknown format variant, other devices in the network may perform one or multiple conversions to adapt the content to the player needs. The first step toward achieving this objective is the development of a language to describe Content Format Variants as well as device encoding, decoding, and conversion capabilities. The next sections describe such a descriptive language.

3.4   DESCRIPTION OF CONTENT FEATURES AND DEVICE CAPABILITIES

There are a few discovery scenarios that home network devices will have to go through during normal operations:

- *Scenario 1: Content Feature Discovery.* A device that implements the CP, MRCVR, and MDEC components is a device capable of browsing content available in the network (CP), receiving content from a source (MRCVR), and decoding content for rendering (MDEC). As a side note, having the CP element together with the receiver/decoder results in a so-called Content Pull Model as defined in Universal Plug and Play (original definition which is no longer in use) (UPnP) [8, 9]. At the time of browsing content from all network servers, this device needs to understand the Coding Format Variant of available media objects to select those that can be decoded by its MDEC component.

- *Scenario 2: Device Feature Discovery.* A device that implements the CP and MSERV components is a device capable of selecting one or more receivers that will receive some particular content item. As a side note, having the CP element together with the source of content results in a so-called Content Push Model as defined in UPnP [8, 9]. At the time of selecting receiving devices, this device needs to understand the Coding Format Variants that the potential receiver could decode.

- *Scenario 3: Device Feature Discovery.* A device that implements the MSTO and MSERV components is a device capable of storing content and making content available to the network. At any time, this device could try to find among the networked devices a converter component (MCONV) to change the format variant of certain media object into another variant that may be stored in anticipation of its use by some other device. At the time of communicating with the MCONV component, both the device and the component need to clearly comprehend the variant conversion combinations that MCONV is capable of performing. The device can then decide if it requests a conversion or not.

In order to implement these baseline scenarios, some protocol or descriptive language is needed to (a) describe the Content Format Variant that has been used to create a binary representation for some content item; (b) describe the decodable Content Format Variants for any device that has decoding capabilities; and (c) describe the input and output Content Format Variants for a device with conversion capabilities.

A method that addresses requirement (a) has been proposed in Refs. 6 and 7 based on the use of a text token known as a "Profile ID." This identifier, included as part of exposed metadata by an MSERV component, describes a particular combination of codec formats and encapsulations. For example, the token "MPEG4_H263_MP4_ P0_L10_AAC_LTP" indicates that the content item matches MPEG4 encoding specifications for H.263 using Profile 0 and Level 10. In addition, this A/V content item carries an audio component conformant to Advanced Audio Coding (AAC) with an Long Term Prediction (LTP) object. Audio and video have been encapsulated using the MP4 file format.

This token is just a text keyword that acts as an index to a particular format variant described in the specifications [6, 7]. The only portion of the string that has some relevance is the first element (MPEG4 in the example above) indicating the family of video coding format. The remaining elements in the string (H263, MP4, etc.) have little relevance because they do not follow any strict syntactic rules. For example, the second element (which is "H.263" in this example) could indicate a completely different feature like in the "MPEG_TS_SD_NA" token. In the latter token, the second element "TS" has been introduced to indicate the multiplexing format "Transport Streams."

The tokens in Refs. 6 and 7 have not been designed to facilitate parsing but only to provide some level of quick recognition to trained humans. Here are a few examples of A/V tokens described in Ref. 7:

- MPEG_PS_NTSC_XAC3
- MPEG4_P2_TS_SP_AAC_ISO
- AVC_TS_MP_SD_MPEG1_L3_ISO
- AVC_MP4_MP_SD_AAC_LTP_MULT7
- WMVHIGH_FULL

The elements in the tokens (MPEG, PS, NTSC, XAC3, MPEG4, P2, TS, etc.) exist mainly to provide some level of human recognition but they have not been designed with the intention of enabling machine parsing and consistent information extraction from individual elements. A trained software engineer will recognize PS as "Program Streams," TS as "Transport Streams," MP as "Main Profile," SD as "Standard Definition," L3 as "Level 3," and so forth. However, these elements do not follow a particular order or some well-defined syntax (except in very restricted subsets; e.g., if one looks only at MPEG-2 or WMV tokens).

Notice for example that the second element sometimes indicates an encapsulation format and sometimes something else. The audio component sometimes is the last element in a token but sometimes it is not. The audio description sometimes needs only one element (XAC3), sometimes two elements (MPEG1 and L3), and sometimes three elements (AAC, LTP, and MULT7). Similarly, the encapsulation component sometime requires one element (PS), and in other cases it requires two (TS and ISO).

The tokens were created with the expectation that a decoding device will extract the token from the object metadata description and then compare the token with the device's own internal list of known tokens. If there is a match, the device will assume that the object is decodable.

Unfortunately, the token-based method does not scale well. The addition of new formats, or new profiles and levels, or new encapsulation methods, or new bandwidth constraints requires creating new tokens. An older device that understands a certain predefined set of tokens will not understand the additional tokens pointing to new information. In many cases, these older devices should have no problem understanding the new variants, but as the tokens do not have strict syntax, older devices will not understand the differences introduced in the new tokens.

For example, consider the token MPEG_PS_NTSC from Ref. 7, which is used to describe content encoded according to the DVD recordable specifications [10].

As defined in Ref. 6, this token corresponds with a combination of audio and video multiplexed using Program Streams. For video, the token describes MPEG-2 encoding with restrictions defined in Ref. 10. For audio, the token describes LPCM, MPEG Layer 2, or AC-3 encoding methods. In the case of AC-3, the maximum audio bit-rate is consistent with Ref. 10 and is defined as 448 Kbps. However, the specifications for AC-3 indicate that the maximum bit-rate goes up to 640 Kbps [11]; a value that is actually in use in the North American digital television system (ATSC).

The difference in bit-rates between the two standards [10] and [11] creates a problem. It is entirely possible to find content items that conform to DVD Recordable [10] but that use an audio bit-rate consistent with the values defined in ATSC [11]. A typical example would be a recording device that captures digital television programs and stores those programs in recordable DVDs. If this device does have the capability of resampling and re-encoding the audio sequences, then chances are high that we would see content that conforms to the MPEG_PS_NTSC profile except that its audio component goes beyond the 448 Kbps limit. How can we describe this new type of content? The solution of course would be to create a new token like MPEG_PS_NTSC_XAC3, which in fact has been proposed for this purpose.

However, the new token will be meaningless for older devices that have not been programmed to understand new tokens. These older devices will be unable to understand that content that comes with the new token has only a minor difference; they will not understand that the only difference is that the audio bit-rate could go up to 640 kbps instead of 480 kbps. Some of the older devices should be able to decode 640 kbps audio, but they will have to discard content items that exhibit the new token simply because these devices cannot understand the meaning of the new token. If instead of, or in addition to the notion of tokens, we could use a declarative language (an XML-based language) with "audio bit-rate" as one of the declared attributes, we would not have this problem.

In addition, the token method would be too cumbersome to resolve requirements (b) and (c) defined above. Consider for example the description of a decoder device capable of decoding North American digital television signals. If this device could in addition decode screen resolutions (horizontal and vertical size of an image) beyond those defined in the standards (a very common occurrence), then some new tokens will have to be created to define those additional resolutions. Every time that a decoder with the ability to decode extra features reaches the market, new tokens will have to be created to understand the new features. Every time that a software decoder upgrades its features through software downloading, it may need a list of newly defined tokens to advertise its new capabilities. Similarly, every time a standard updates the features for some of its media definitions, newly defined tokens may be required to identify those differences. Again, older devices will be unable to recognize the new tokens even if in practice, some of those devices are able to decode the new formats. These scalability problems seriously limit the efficiency of the token method described in Refs. 6 and 7.

A better approach seems to be to use a descriptive language based on XML to express the features of Content Format Variants, which will then be used to describe properties of media objects (requirement [a]), or device decoding/conversion capabilities (requirements [b] and [c]). XML is flexible enough to incorporate additional features for future usability (e.g., in the future, we may want to add content protection information in addition to encoding and encapsulation features). In the next sections, we discuss the construction of a meta-language that we call here MXDL for "Media eXchange Description Language."

3.5 MEDIA EXCHANGE DESCRIPTION LANGUAGE

In this section, we introduce and propose a description language (MXDL) that could be used to model the diversification of media format variants as a means to solve the diversification problem. We approach this subject as a tutorial, avoiding the complexities arising from the need to define strict schema declarations, name spaces, document formats, possible Simple Object Access Protocol (original definition which is no longer in use) (SOAP) bindings, and so forth. A full-featured specification should be available in the future.

A simplified structure of MXDL messages is shown in Figure 3.4. A single message (or document) needs to have Primitives and Compounds. The term *primitives*, in this context, is used to describe audio, video, and stream composition features. A primitive declaration that uses the element <Video> indicates the coding format for video including possibly profiles and levels, restrictions on resolutions, and so forth. A primitive declaration that uses the element <Audio> indicates the coding format for audio including possibly bandwidth constraints, sampling rate constraints, and so forth. A primitive declaration that uses the element <StreamOp> indicates a stream operation like encapsulation, multiplexing, payload formatting, encryption, and so forth. An MXDL message has only one section for primitives, but it could have as many primitive components as necessary.

The section on Compounds is used to define compositions of primitives that identify the properties of a media object or the capabilities of a device. For example, the primitives may indicate support for MP3, AC-3, and MPEG-2 video, MP3 file format,

```
<MXDL>
        <Primitives>
            . . .
            . . .
        </Primitives>

        <Compounds>
            . . .
            . . .
        </Compounds>
</MXDL>
```

FIGURE 3.4   Main sections of an MXDL message.

and Transport Streams. The section on Compounds could indicate that one valid A/V combination is MPEG-2 video with AC-3 audio encapsulated using Transport Streams. In addition, the section on Compounds could indicate that MP3 encoded audio encapsulated in the MP3 file format is another Variant that can be composed using the described primitives. A decoder that uses this message to advertise its decoding capabilities would be declaring its ability to decode audio signals in the form of MP3 and A/V signals in the form of MPEG-2 video with AC-3 audio.

In practice, MXDL will use registered name spaces and controlled lists of element and attribute values. For example, the known codecs like MPEG-2, AAC, WMV, and so forth, will exist as part of the initial MXDL definitions. If in the future, a newly invented codec becomes popular, a new version of MXDL will have to incorporate support for the new codec. Older devices would be unable to understand the new version of MXDL, which means that MXDL solves only partially the diversification problem. MXDL solves the diversification problem arising from well-known constraints and extensions imposed on the existing list of known codec algorithms. For example, if in the future some combination of WMV with AAC becomes popular, MXDL would be able to describe the new variants. Devices designed to understand MXDL will be capable of understanding the new variants using conventional MXDL descriptions.

3.5.1 MXDL Media Object Descriptions

This subsection describes the use of MXDL for media object descriptions. When some device in the network searches for media objects, it needs to know the particular Content Format Variant that has been used to encode, encapsulate, and possibly encrypt a media object. MXDL could be used to advertise these features.

Figure 3.5 shows an MXDL message used to describe an object's Content Format Variant. In this example, the object includes audio and video components. An examination of the <Video> primitive reveals that video is encoded using the MPEG-2 standard using Main Profile at High Level. The encoding of video has been done in such a way that it follows the ATSC standard and uses a resolution of 1280 × 720. Note that, in this case, declaring the resolution information is superfluous because the ATSC standard includes a resolution of 1280 × 720 as part of the specification. However, the more information an author provides, the easier the decoding process becomes.

From Figure 3.5, an examination of the <Audio> primitive reveals that the audio component for this object conforms to AC-3 with a bit-rate restriction of 448 Kbps. This restriction comes handy because it signals some potential DVD players in the network that in terms of bandwidth, the audio signal is compatible with their decoding capabilities. According to the DVD specifications, the maximum bit-rate for AC-3 audio is precisely 448 Kbps.

The <StreamOp> primitive element indicates an operation toward composing a stream. In this case, the operation consists in creating a byte sequence conformant to

```
<MXDL>
   <Primitives>
      <Video Type="MPEG2" Profile="Main" Level="High" Id="V1" >
         <Restrictions>
            <Resolution Type="pixel"> 1280-720 </Resolution>
            <Standard> ATSC </Standard>
         </Restrictions>
      </Video>
      <Audio Type="AC-3" Id="A1" >
         <Restrictions>
            <BitRate Type="kbps"> 448 </BitRate>
         </Restrictions>
      </Audio>
      <StreamOp Type="MPEG2-TS" Id="E1">
         <Restrictions>
            <Standard> ATSC </Standard>
         </Restrictions>
      </StreamOp>
   </Primitives>

   <Compounds>
      <Instance>
         <Variant>
            <StreamConfig   Ref="E1">
               <AudioConfig   Ref="A1" />
               <VideoConfig   Ref="V1" />
            </StreamConfig>
         </Variant>
      </Instance>
   </Compounds>
</MXDL>
```

FIGURE 3.5 Example MXDL message describing the Content Format Variant for a media object in a home network.

the MPEG-2 Transport Stream specifications as defined in Ref. 2. Notice that in the example of Figure 3.5, the TS description includes an additional restriction: the Transport Stream matches ATSC specifications. In ATSC, for example, there are restrictions for aligning MPEG-2 PES packets with Transport Stream packets [12]. Some decoders tailored specifically to the ATSC environment need this information before proceeding to decode some signal. Other decoders do not care about the particular ATSC restrictions and should be able to decode the signals ignoring the defined restrictions.

Finally, the element <Compounds> has a single child element called <Instance>, which indicates that the compound operations will result in a single object description instance. There is only one <Instance> element as a child of the <Compounds> element.

An <Instance> element can have one or more <Variant> children. In most cases, it will only have one <Variant> element because an object only accepts one Variant. However, there are a few cases where an object could be represented by two or more variants simultaneously. For example, consider the case of an A/V object that exists as an MP4 file in a server. The server could expose the MP4 file to the network, but in some cases the server could also expose the object as separate elementary streams for Real-time Transport Protocol (RTP) transfer. In this case, it is clear that this object can be represented using two alternate Variants.

In the example of Figure 3.5, there is only one Variant, which is the typical case for most usages. The Variant in Figure 3.5 explains that the object has been created using the referenced Audio and Video primitives and configuring a stream according to the instructions defined by the StringOp primitive.

In summary, the message shown in Figure 3.5 conveys the following information: a media object that includes audio and video components, encapsulated in Transport Streams following ATSC provisions. The audio component is AC-3 with bit-rate constrained to 448 Kbps (which actually matches the audio bit-rate requirements of DVD applications). The video component is MPEG-2 with constraints defined in the ATSC standard; constraints that include resolutions, aspect ratios, frame rates, progressive versus interlaced, and so forth.

A typical A/V object in MXDL needs only the video, audio, and the stream operation primitives to create a single Variant (which is the case shown in the example of Fig. 3.5). More complex media objects, however, can also be represented by MXDL. Here we describe briefly four of those cases.

Case 1: Objects with Multiple Components. It is entirely possible (and becoming more and more common) to have Audio or A/V objects that carry multiple components, each using a different coding algorithm. For example, an MP4 file could reveal video encoded using H.264, and two audio components, one using High-Efficiency AAC (HE-AAC), and the other using Adaptive Multirate (AMR). In MXDL, this case is easily represented using one video primitive, two audio primitives (one for HE-AAC and one for AMR), and one stream operation primitive for creating the MP4 file.

Case 2: Objects Prepared as Multiple Streams. In scenarios that use RTP streaming, the individual components (audio or video) are often streamed separately as Elementary Streams following well-defined payload formats. Consider for example an object that for storage exists as an MP4 file carrying H.264 encoded video and AAC audio. For RTP streaming, this object is often available as two separate elementary streams. The AAC audio stream can be configured following the payload format defined in Ref. 13, whereas the H.264 video stream could follow the payload format defined in Ref. 14. If we assume that the MP4 file is not available over the network, and only the RTP-based streams could be transferred, then an MXDL representation for this object will show a <Variant> element with two <StreamConfig> children elements (one for the audio stream and one for the video stream). The example bellow shows this case. In the example, the <StreamOp> primitive referenced by "op1" will describe the RTP payload format for configuring the audio stream. Similarly, the <StreamOp> primitive referenced by "op2" will describe the RTP payload format for configuring the video stream:

```
<Variant>
  <StreamConfig Ref="op1">
     <AudioConfig Ref="a1"/>
  </StreamConfig>
  <StreamConfig Ref="op2">
     <VideoConfig Ref="v1"/>
  </StreamConfig>
</Variant>
```

Case 3: Objects with Multiple Variants. The notion that an object can be represented using two or more variants has been discussed already in this document. A typical example would be an A/V object that could be exposed as an MP4 file object or as separate audio and video elementary streams for RTP streaming. An MXDL representation for this object will show one <Instance> element with two <Variant> children.

Case 4: Objects that Require Multiple Stream Operations. All the examples mentioned so far show audio and video primitives combined using some stream operation (e.g., the stream operation that generates a Transport Stream in Fig. 3.5). In many cases, after creating a multiplexed stream, additional stream operations are required in order to carry an object across the network. The most typical example is encryption. MXDL allows recursive usage of the <StreamConfig> element to enable cascaded operations. The following example shows how the recursive model works. In the example, audio and video are multiplexed using Transport Streams defined by the <StreamOp> primitive with "op-ts" identifier. The resulting stream is then encrypted using some encryption algorithm declared previously in the <StreamOp> primitive with the "op-enc" identifier:

```
<StreamConfig Ref="op-enc">
   <StreamConfig Ref="op-ts">
      <AudioConfig Ref="a1"/>
      <VideoConfig Ref="v1f"/>
   </StreamConfig>
</StreamConfig>
```

3.5.2   MXDL Device Capability Descriptions

In order to understand how MXDL could be used to describe device capabilities, we need to introduce first a systems model for networked media devices. From the

FIGURE 3.6 The systems model for describing media processing capabilities of a device.

perspective of media home networking, any media-processing device can be represented as a system that takes multiple inputs and generates multiple outputs. Figure 3.6 illustrates this concept. The inputs in this case are Content Format Variants, and similarly the outputs also correspond with Content Format Variants.

A device with no outputs represents a pure sink device. A sink device consumes content but never generates content as an output. A DTV monitor that implements an MRCVR and an MDEC receives content from the network for decoding and rendering only. This device would be a perfect example of a pure sink device. In this case, this device may have one or more inputs—the Content Format Variants that it can decode—and zero outputs.

Similarly, a device with no inputs represents a pure content source that creates content for the network but it is unable to receive content for conversion or pass-through. A DTV set-top box that implements MCAP and MSERV becomes a pure content source if it is unable to receive content from other devices in the network and the only task it performs is serving digital television signals to the network. A digital camera that exposes JPEG files to the network but that is unable to receive files from the network would be a second example of a pure source device. Using the systems model of Figure 3.6, a pure source device will have zero inputs and one or more outputs. The outputs correspond with the Content Format Variants that could be exposed to the network.

Some devices receive content and also output content. One example would be a content conversion device (which implements the MCONV component) that changes content from one format variant to another. If any of N variants at the input can be converted into any of M variants as the output, then the system representation for this device will have N inputs and M outputs.

Often, conversion devices will actually work as a multiple-system device. This means that a conversion device will convert any of N_1 variants into any of M_1 variants. In addition, the same conversion device will be capable of converting from any of N_2 variants into any of M_2 variants, and so on. For example, a device capable of A/V format conversion could implement two systems. The first system could convert from a few of the MPEG-2 variants into a few of the WMV variants. The second system could convert from a few of the H.263 variants into a few of the H.264 variants.

The systems-based abstraction illustrated in Figure 3.6 is incorporated in MXDL through the use of a "Systems" element under the "Compounds" section.

```
<MXDL>
    <Primitives>
        . . .
        . . .
    </Primitives>

    <Compounds>
        <System>
            <Input>
                <Variant ... ... ..... />
                <Variant  ... ... ..... />
                <Variant  ... ... ..... />
            </Input>
            <Output>
                <Variant ... ... ..... />
                <Variant  ... ... ..... />
            </Ouptut
        </System>
        <System>
            <Input>
                <Variant  ... ... ..... />
            </Input>
        </System>
    </Compounds>
</MXDL>
```

FIGURE 3.7 Simplified MXDL message for describing media processing capabilities of networked devices.

Figure 3.7 shows a simplified structure for a message describing device capabilities using the systems Inputs and Outputs (I/O) model. Upon detection of a "Systems" element, the parser identifies this message as a device capability description and not as a media object description (which has an "Instance" element instead). In summary, <System> elements always describe device capabilities in accordance with the systems model introduced in this document. On the other hand, <Instance> elements always describe media object encoding features.

Notice in Figure 3.7 that the syntax defined for MXDL accommodates the option of implementing multiple systems. In the example shown in Figure 3.7, two systems have been declared for a certain device in the network. The first one takes three Content Format Variants as the input and generates two variants at the output. The second system describes pure sink behavior as it admits one input variant but generates no output variants.

Figure 3.8 shows an actual example of declaring device capabilities using MXDL. The device in the example has been modeled as a three-system device. Two of the systems have inputs and outputs indicating that this device can do format variant conversions in two different ways. The third system only shows output elements indicating that this device can also act as a sink device (e.g., as a device capable of rendering certain types of content variants).

```
<MXDL>
    <Primitives>
        <Video Type="MPEG2" Profile="Main" Level="High" Id="V1" >
           <Restrictions>
               <Standard> ATSC </Standard>
           </Restrictions>
        </Video>

        <Video Type="VC1" Profile="Advanced" Level="2" Id="V2" />

        <Audio Type="AC-3" Id="A1" />

        <Audio Type="WMA" Id="A2" />

        <Audio Type="MP3"  Id="A3" >
           <Restrictions>
               <BitRate Type="kbps"> 192 </BitRate>
           </Restrictions>

        <StreamOp Type="MPEG2-TS" Id="E1">
           <Restrictions>
               <Standard> ATSC </Standard>
           </Restrictions>
        </StreamOp>

        <StreamOp Type="ASF" Id="E2" />

        <StreamOp Type="MP3FF" Id="E3" />

    </Primitives>

    <Compounds>
        <System>
           <Input>
              <Variant>
                 <StreamConfig   Ref="E1">
                     <AudioConfig   Ref="A1" />
                     <VideoConfig   Ref="V1" />
                 </StreamConfig>
              </Variant>
           </Input>
           <Output>
              <Variant>
                 <StreamConfig   Ref="E2">
                     <AudioConfig   Ref="A2" />
                     <VideoConfig   Ref="V2" />
                 </StreamConfig>
              </Variant>
           </Ouptut>
        </System>

        <System>
           <Input>
              <Variant>
                 <StreamConfig   Ref="E2">
                     <AudioConfig   Ref="A2" />
                 </StreamConfig>
              </Variant>
           </Input>
```

FIGURE 3.8 Example MXDL message describing the media processing capabilities of a device connected to the home network.

```
        <Output>
            <Variant>
                <StreamConfig    Ref="E3">
                    <AudioConfig    Ref="A3" />
                </StreamConfig>
            </Variant>
        </Output>
    </System>

    <System>
        <Output>
            <Variant>
                <StreamConfig    Ref="E3">
                    <AudioConfig    Ref="A3" />
                </StreamConfig>
            </Variant>

            <Variant>
                <StreamConfig    Ref="E2">
                    <AudioConfig    Ref="A2" />
                </StreamConfig>
            </Variant>

        </Output>
    </System>
    </Compounds>
</MXDL>
```

FIGURE 3.8   *Continued.*

If one follows the attribute identifiers "Id" in the primitives and matches them with the attribute reference identifiers "Ref" in the compounds, it is easy to recognize that the first system usage in Figure 3.8 corresponds with the capability of converting digital television content into PC-friendly content. The format for the digital television content consists of MPEG-2 video and AC-3 audio encapsulated using Transport Streams with constraints defined in ATSC specifications [12]. The PC-friendly format at the output consists of video encoded using VC-1 and audio encoded using Windows Media Audio (WMA), encapsulated in Advanced Systems Format (ASF) files. This is the typical configuration for A/V content when using the Windows Media Player software in PCs. Notice that VC-1 [5] is the standardized version of Microsoft's well-known Windows Media Video coding format.

The second system usage in the example of Figure 3.8 indicates that this device can also convert WMA files into MP3 files, probably for use by some devices unable to play the WMA format. The conversion process outputs MP3 files with a maximum audio bit-rate of 192 kbps, a value that is typical in current implementations. Also notice that the MP3 output conforms to the "MP3 file format," which normally includes ID3 tags for title, artist, genre, and other relevant information.

The third and final system usage in the example of Figure 3.8 indicates that this device can also act as a sink device for the purpose of rendering content. In this case, when acting as a rendering device, it can only play audio streams of the WMA and MP3 variants defined in the MXDL message.

3.6 CONCLUSIONS

A plausible outcome for the current trend in wireless and wired home networking seems to be that in the near future, most media-centered devices will be capable of exchanging high-quality media content with other devices in the home network. We believe that this plausible outcome actually shifts the focus of networking from office-related applications (sharing printers, sharing Internet access) into a content-oriented home network. This shift in perspective obviously brings new challenges at many levels; from physical infrastructure devices that will need to cope with high bandwidth requirements to intelligent software agents that will perform content format conversions to maximize the availability and reproducibility of content within the network.

This chapter describes the problems related to the diversification of media formats that occurs in practice with the aging of media standards. We show that the normal evolution over time of digital media formats leads usually to format variants customized to meet the requirements of specific applications. For example, the MPEG-2 coding and transport standard has evolved into variants that satisfy the requirement of DVD, digital television (ATSC, DVB, and ARIB), and several other similarly focused applications. Although these variants serve very well those customized applications, they become a serious problem when they have to be integrated and processed in a home network. An expectation of the users of content home networks relates to the ability to render content in any sink device regardless of the source of the content. The existence of a multiplicity of format variants in the home becomes an obstacle to fulfill this expectation. One way to minimize this problem is to enable software agents that will do content conversion in real time or off-line each time that a disparity occurs between content types and rendering devices.

Given the diversity of media format variants in the field, some systematic description of their features is necessary if we want devices to understand the type of content that is available. This systematic description could allow devices to recognize *a priori* if they can process or not certain media objects, and it could allow devices to describe their media processing capabilities to other devices in the home network. In this chapter, an XML-based language is used to construct this systematic description (the Media eXchange Description Language, or MXDL).

The language described in this chapter is only a first step toward solving this complex problem. In addition, new media negotiation protocols will be necessary to provide a complete solution to the problem. These media negotiation protocols will allow a server to query for the MXDL messages associated with certain objects and will allow a receiving device to request conversion services when the object's MXDL description does not match its own decoding capabilities. Similarly, a conversion device will use a query protocol to understand the decoding capabilities and preferences of the receiving device before deciding on a particular format conversion algorithm.

Eventually, it is possible that all the pieces will be in place to enable transparent operation of the content home network. When this happens, content rendering in home networks should approximate the transparent and simple plug-and-play experience that users would like to have. Content will enter the home network from multiple sources using multiple format variants, but it will be transparently processed to maximize its quality and availability for all devices within the home network.

REFERENCES

1. E. Heredia, "Media Exchange Protocols for Home Networks." Proceedings of IWS 2005/ WPMC'05, September 2005.
2. "Generic Coding of Moving Pictures and Associated Audio Information: Systems." ISO/ IEC 13818-1, Specifications from the International Organization for Standardization.
3. "Generic Coding of Moving Pictures and Associated Audio Information: Video." ISO/ IEC 13818-2, Specifications from the International Organization for Standardization.
4. "Generic Coding of Moving Pictures and Associated Audio Information: Audio." ISO/ IEC 13818-3, International Organization for Standardization.
5. "Standard for Television: VC-1 Compressed Video Bitstream Format and Decoding Process." SMPTE 421M, Specifications from the Society of Motion Picture and Television Engineers (SMPTE), 2006.
6. "DLNA Networked Device Interoperability Guidelines Expanded—Volume 1: Architecture and Protocols." Specifications from the Digital Living Network Alliance (DLNA), March 2006.
7. "DLNA Networked Device Interoperability Guidelines Expanded—Volume 2: Media Format Profiles." Specifications from the Digital Living Network Alliance (DLNA), March 2006.
8. "UPnP Device Architecture 1.0 Version 1.0.1." Specifications from the UPnP Forum, December 2003.
9. "UPnP AV Architecture: 0.83." Specifications from the UPnP Forum, June 12, 2002.
10. "DVD Specifications for DVD-RAM/DVD-RW/DVD-R for General Disks Part 3 Video Recording." Specifications from the DVD Forum, Version 1.1, 2001.
11. "Digital Audio Compression Standard, Rev. B." ATSC A/52B, Specifications from the Advanced Television Systems Committee (ATSC), 2005.
12. "Digital Television Standard, Rev D with Amendment 1." ATSC A/53D, Specifications from the Advanced Television Systems Committee (ATSC), 2005.
13. "RFC 3640, RTP Payload Format for Transport of MPEG-4 Elementary Streams." IETF, November 2003.
14. "RFC 3984, RTP Payload Format for H.264 Video." IETF, January 2005.

None of the improvements to the standards described in this chapter are present in the versions of Windows available at the time this book was published (Windows XP, Windows 2000, Windows Vista), and Microsoft has no obligation to market, sell or otherwise distribute the described improvements, either alone or in any Microsoft product or service.

4

MOBILE DEVICE CONNECTIVITY IN HOME NETWORKS

MIKA SAARANEN AND DIMITRIS KALOFONOS

Networking at home has traditionally meant connecting home PCs to the Internet through a modem line or a broadband connection. Local networks such as Ethernet have been very rare in the home environment. Home networking has its roots in building automation applications, with technologies like LonMark [1] or X.10 [2], but these have not been widely used in private homes, although many new buildings do incorporate this kind of technology. This traditional image is now being changed with the emergence of digital media consumed natively on various home devices and with the current boost on Wireless Local Area Networks (WLANs) at home and in public places.

Despite the focus being on providing connectivity over wide-area networks, the Internet community has defined and deployed almost all necessary technology components needed for home networking and consuming media. For instance, Transmission Control Protocol/Internet Protocol (TCP/IP), HyperText Transmission Protocol (HTTP), and Real Time Protocol (RTP) have been defined in Internet Engineering Task Force (IETF) and deployed also on other networks like General Packet Radio Service (GPRS) or Universal Mobile Telecommunications System (UMTS). In addition, media formats like Joint Photographic Experts Group (JPEG) and Moving Picture Experts Group (MPEG)-1 Layer 3 (more commonly known as MP3) are widely adopted by the Internet community.

When these technologies are combined with WLANs and Ethernet, we have the seeds of an emerging new era of home networking. From the manufacturers' point of view, this Internet technology–based home networking appears appealing because it enables new product concepts. First, completely new kinds of devices for home media consumption and storage can be made, such as media servers that

Technologies for Home Networking. Edited by Sudhir Dixit and Ramjee Prasad
Copyright © 2008 John Wiley & Sons, Inc.

can store hundreds of hours of movies or thousands of individual songs. On the other hand, digital content can be stored and served from home PCs after it has been downloaded from the Internet, as some already existing products already demonstrate. Second, networked devices may exclude local user interface allowing both for smaller devices and, especially, for lowering the manufacturing cost of the devices. In this scenario, a networked remote controller, possibly provided by a different vendor, would provide the user interface. Furthermore, common networking technologies promise a reduction on the amount of distinct wiring necessary, thus making home networking more feasible for consumers. Remote control of home media devices may also extend to mobile terminals capable of Internet connectivity, even if they are used in remote locations.

When consumers start deploying networked home appliances, usability and ensured interoperability between home appliances is of prime importance. Although usability is mostly based on manufacturers' ability of designing good products, the ensured interoperability requires standardization or industry-approved guidelines. There have been many efforts in standardizing a home architecture, such as the Home Audio Video interoperability (HAVi) [3] and the Digital Living Network Alliance (DLNA) [4]. For instance, these two forums define a complete set of technologies providing all pieces from communication to applications and from zero-configuration to advanced control of home devices. These forums have applied different approaches to the architecture, making their approaches more or less applicable to different audiences.

In this chapter, we focus on home networking including local and remote connectivity focusing mostly on mobile devices. We briefly review related work on home connectivity and some of the most interesting basic use cases. We discuss the challenges that face both users and manufacturers. We argue that the most important challenges are on usability and automatic configuration of the home network and its devices. We also present a home networking architecture based on Internet and IEEE 802 technologies and a review of many alternative connectivity methods. We also discuss remote connectivity and present two potential technologies for mobile devices accessing home appliances remotely.

This chapter is organized as follows: Section 4.1 presents a brief overview of related work; Section 4.2 introduces some basic use cases for home connectivity; and Section 4.3 introduces the challenges of building solutions for home networking. Section 4.4 introduces architecture and potential technologies for local and remote home connectivity. Finally, the conclusions of this work are discussed in Section 4.5.

4.1 RELATED WORK

Related work in this area can roughly be divided into two groups: standardization and academic research. Furthermore, there are many exciting products emerging in the industry, but in this chapter we will not refer to commercial product efforts. The two first standardization forums presented here (HAVi and DLNA) have defined a decentralized home network that is mainly targeted for home entertainment applications.

The HaVi [3] approach to home networking is based on IEEE 1394, which provides high data rates with guaranteed Quality of Service (QoS). HAVi defines a full architecture with hot plug-and-play features. Device architecture allows installing Java-based software (SW) modules for the provision of device control. HAVi provides non-IP networking but facilitates gateway solutions that allow Internet connectivity. DLNA [4] has taken an alternative approach, basing its technology choices on protocols and formats used mainly in the Internet community. Device discovery and control is based on Universal-Plug-and-Play (UPnP) [5]. DLNA aims at creating interoperability guidelines based on existing standards rather than defining new technologies. The goal is to instruct device manufacturers to use DLNA-defined technologies in an interoperable way. In this chapter, we will present an architecture that is compatible with this DLNA approach while describing also some technologies beyond the current DLNA choices. The DLNA guidelines are currently focusing on local home networking and providing media devices interoperability. Although its upcoming interoperability guidelines (DLNA v1.5) will also address mobile devices and their operation at home as one interoperable but distinct device class, remote home connectivity has not yet been in the scope of DLNA. In this chapter, we are presenting potential options to implement mobile device connectivity.

An alternative approach, which can be characterized as centralized, has been taken by the Open Services Gateway Initiative (OSGi) [6]. OSGi is a versatile framework that provides application programming interfaces (APIs) and Java execution environment for building integrated home networks and adaptation functions between various other solutions like UPnP-controlled devices. In this approach, various networking and service provision environments are connected through services that the OSGi gateway provides. If the OSGi gateway and its services are well implemented, it may provide valuable additional services without invalidating existing networked services.

In the academic research literature, most advanced solutions are part of the *pervasive computing* area of research. Advanced solutions are proposed, including determining the location or intent of the people and acting based on observed behaviors. A well-known example in this area is Project Aura from Carnegie-Mellon University (CMU) [7]. The work in Ref. 8 presents a proposal that provides a ubiquitous computing environment with location-based services at home. The implementation uses mainly standard protocols such as Service Location Protocol (SLP) [9] and Session Initiation Protocol (SIP) [10] for controlling and discovering home appliances.

Another direction of research focuses on building middleware solutions that aim to solve problems, such as zero-configuration, usability, and interconnecting noncompatible technologies. In Ref. 11, the authors present a unified home services interface behind which real devices are hidden and meta-devices that can combine several individual devices. An OSGi-based home architecture is presented in Ref. 12.

4.2 BASIC HOME USE CASES

In this chapter, we focus on home entertainment–related use cases and on the local and remote access connectivity issues related to these use cases. In the rest of this

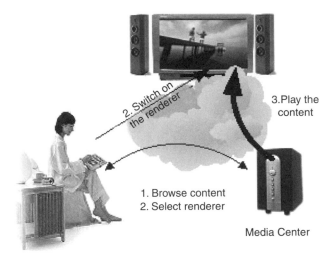

2. Switch on the renderer

3.Play the content

1. Browse content
2. Select renderer

Media Center

FIGURE 4.1   Use case for locally watching a movie.

section, we present four main use cases, from which more detailed use cases could be derived.

Maybe the most frequent use case in the home environment is to consume media content from a local or external source. Media content can be, for instance, your favorite TV show or listening CDs or MP3s. Figure 4.1 shows a typical case where a user is watching a prerecorded movie from a media server. In this case, the user will activate the TV set and the media server, browse, and select the desired movie and start playing the movie at the selected TV set. The media server can be, for example, a media PC, a DVD player, or a set-top box. After watching the movie, the user may select a new program or turn off the devices. This scheme is also valid for listening to radio or CDs or MP3s from stereo equipment.

Related to the previous case, in many cases people desire to record broadcasted programs for later consumption, for example, when the weekly transmission of some program happens at an inconvenient time. Actual programming is usually done while in close proximity to the recording device, but the remote access case shown in Figure 4.2 is also interesting, as the mobile lifestyle is getting more popular. The basic use scenario, either in the remote or the local programming case, would be first activating the remote controller. In the local case, this may just mean picking the remote controller, but in the remote case the user would have to start a home access application and potentially accept usage of a cellular or public WLAN network. The next phase would be finding out when and in which channel the desired program shall run. In an advanced scenario, the home recording device would show program information from an Electronic Programming Guide (EPG) and the user could just select the program. However, the current way of operation (i.e., just entering the starting and ending times and the channel numbers) is expected to remain as a backup method. After the programming has been completed, the user may also want to see the list of his or her scheduled recordings.

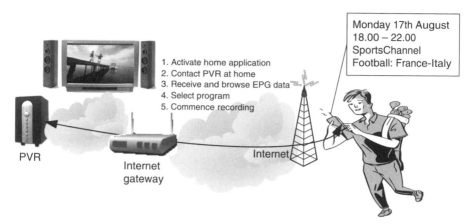

FIGURE 4.2 Use case for remote Program Video Recorder (PVR) programming.

In the current MP3 player boom, managing and synchronizing a music collection will have a significant role. In particular, smaller music players may store only a fraction of the whole music collection, therefore creating the need to update the collection while inside or outside the home. As WLAN connectivity will provide almost free access, such a mobile update will be feasible also at a reasonable cost.

Currently, sharing of multimedia content is getting more and more popular. This would be the fourth type of use case, where sharing media content between friends will have a significant role. In relation to home usage, the typical scenario would be exchanging selected songs with your friends visiting your home. This sharing presents legal challenges because the copyrights of the content owners should be protected, but at the same time legal sharing should not be limited.

There are many alternative use cases that can be derived from the examples we described and also totally unrelated to these. Currently, the digital networked home is taking its first steps and therefore new ways of exploiting this technology will emerge in the future.

4.3 HOME NETWORKING CHALLENGES

There are a number of technical challenges that have to be addressed in order for home networks to be deployed successfully and be in a position to enable this multitude of new use cases. The main difference between home networks and traditional networks in the corporate world is that the persons setting up and maintaining them are everyday consumers, with little or no knowledge of networking technologies. Also, the deployment of these home networks is happening in a completely decentralized way, without coordination among the consumers located in the same area, in contrast with professionally installed networks where there is a significant effort in planning prior to deployment of potentially interfering networks. The above factors, combined with the market pressure to minimize the cost of home networking products, create some unique challenges in research aiming at home networking.

In general, it is expected that home networks will be set up and managed by everyday consumers and not by trained professionals. Therefore, the networking technologies employed must enable home networks to be *self-configurable* as much as possible, with minimal or no intervention from the consumers themselves. The required approach puts the burden on product and protocol design rather than network administrators, as opposed to the traditional approach in networking. The home network architecture should include autoconfiguration protocols (e.g., [13], [14]) that will enable a "buy-plug-and-play" experience for consumers. Besides the required protocols, the product designs should provide a user interface that should hide the technical network complexity from the users as much as possible and interact with them only through intuitive mechanisms (e.g., through interactive "wizards," real-world intuitions such as "touch" and "point-and-click"). The home network architecture should support incremental network deployment, enabling the easy and seamless integration of new network nodes as consumers buy and bring home new products. Finally, the product design should enable nonexperts to easily make administrative changes to existing home networks to accommodate new users, new services, and new usages of home devices.

Besides the challenge of enabling self-configuring home networks that are appropriate to be set up and maintained by nonexpert consumers, it is equally important that the home networks are designed to be as *self-healing* as possible. Because home networks are deployed by nonexperts, with little or no planning and no coordination with other consumers, it is very probable that they will be prone to erratic behavior. Examples of problems that may be encountered are radio interference and cross-talk, inadequate bandwidth, excessive round-trip delays, network congestion, addressing conflicts, uneven wireless coverage, inappropriate timer time-outs, as well as a range of other problems. The consumers would be unable to pinpoint the source of all these problems and are likely to attribute the problems to the home products themselves. This can lead to a very frustrating user experience and bad marketing perception for the manufacturers of home devices. Creating self-healing home networks requires both new protocols and middleware, as well as an approach to product and application design that anticipates erratic network behavior and handles it gracefully. For example, disconnections should not stall node resources and require rebooting by users, but rather products should be designed to autoresume where possible or exit gracefully by providing easy-to-understand error explanations to users. Finally, manufacturers would have to develop new network monitoring and debugging tools, which would be easy-to-use by nonexpert consumers and would suggest to them potential fixes whenever problems occur. Alternatively, new tools and services would have to be provided by third-party technical support companies under contract by the consumers, which will monitor, detect, and correct home network problems remotely, with minimal or no intervention by the consumers themselves.

Another important challenge in the design of home networks is *security*. Many of the component networking technologies (e.g., 802.11 [15], HomePlug [16], cellular networks for remote access) are based on sharing a transmission medium that is accessible by many other users. This creates a potential vulnerability for home networks that may be exploited by malicious users with a devastating effect on the usability of the home networks. In a professionally administered private network environment,

administrators take several countermeasures to thwart attacks and allow only authorized access to the nodes and services in the private network. However, such sophis-
tication cannot be assumed in a home network environment. The home architecture
should include enough security mechanisms at the link level (e.g., Wired Equivalent
Privacy (WEP) [15], Wi-Fi Protected Access (WPA) [17], 802.1X [18], 802.11i [19],
Bluetooth Security [20]), network level (e.g., IPsec [21]), transport level (e.g.,
Transport Layer Security (TLS) [22], HTTP Secure (HTTPS) [23]), and application
level (e.g., UPnP Security [24]). Equally importantly, the applications and products
should be designed in a way to allow easy configuration of security because otherwise
consumers will not use it. The example of the proliferation of 802.11-based home
networks, many of which are even now used without security enabled, gives a colorful
illustration of the magnitude of this problem. Fortunately, the industry has realized the
problem and there are important efforts under way to address it. For example, the Wi-Fi
Alliance has formed a working group to create an interoperable solution that will enable
the easy setup of security settings of Wi-Fi compliant 802.11 access points. Finally, as
in the case of regular network problems, new tools and services would have to be
developed that would detect and address the inevitable malicious attacks in home networks. Again, this should happen with minimal or no intervention by the consumers
themselves and could be performed by third-party companies under contract.

Remote access to home networks is also expected to present a range of serious
challenges both to product designers and to consumers themselves. The users
would like to access the devices and services of their home networks remotely
while enjoying an experience as close as possible to that of being physically
located at home. However, the location and method of remote network attachment
can create vastly different connection characteristics when accessing the home
remotely. Example of characteristics that vary widely include the bandwidth and
round-trip delay (e.g., GPRS vs. DSL vs. 802.11), cost (e.g., cellular vs. 802.11),
security requirements (e.g., access from a corporate intranet vs. access from a Café
WLAN hot spot), and capabilities of the devices used for remote access (e.g.,
desktop vs. laptop vs. mobile phone). It is an important challenge to design
systems and products that create a user experience as uniform as possible and that
allow the user to select the best available remote access method based on his or
her desired task (e.g., remote recording programming vs. content streaming),
context (e.g., location), and preferences (e.g., connection cost profile). Finally,
other challenges related to remote access include the use of dynamic IP addresses
in typical broadband connections, as well as traversing firewalls and network
address translation (NATs) that effectively prevent outside connectivity.

Furthermore, mobile devices face particular challenges when connecting to home
networks. Specifically, mobile devices face more frequent disconnections as users
move in and out of range of local access points and experience signal fading or interference when users roam while remotely accessing their home networks. Also,
mobile devices have limited resources such as computational power, battery life,
and screen size, all of which pose extra networking requirements when accessing
the home networks locally or remotely. Finally, it is mobile devices that are most
commonly used when visiting other home networks. Enabling the "visitor scenario"
poses more challenges both for users and manufacturers, both from the home network

architecture and system design perspective, as well as from the perspective of usability and ease of use.

Finally, it is important to note that even if research in the home networking area produces proposals that solve most of the above technical issues, there are other nontechnical challenges that need to be addressed before home networks can become pervasive. Specifically, two important issues are interoperability and deployment aspects of the necessary technology components. On one hand, interoperability of devices is crucial because home networks will most likely be composed of a multitude of devices from many different vendors. Manufacturers of stand-alone devices often base their designs on proprietary technologies. However, this approach would only allow networked home devices of the same manufacturer to work together, an unlikely scenario as this would severely limit the choices available to consumers. On the other hand, deployment issues of the various technology components can significantly affect the success of home network products. Therefore, proposals should consider the deployment schedules of the necessary technologies and not make unrealistic assumptions about their availability. For example, products that rely on IPv6 or Near-Field Communications (NFC) technologies should also implement alternative technologies (e.g., dual IPv4/IPv6 stack, Infrared-IrDA interfaces), as it is unlikely that all devices in home networks will feature these new technologies in the immediate future.

4.4 ARCHITECTURE AND TECHNOLOGIES FOR LOCAL AND REMOTE HOME CONNECTIVITY

4.4.1 Overview of Home Connectivity Architecture

As mentioned earlier, there are many alternatives for the home networking architecture. Whereas many proposals suggest using a centralized architecture that provides a unified service framework, the trend seems to be moving toward a decentralized approach built on IEEE 802 LAN protocol-family and Internet-based technologies. This approach allows home networks to be built gradually based on individual needs. Also, as these technologies are mostly mainstream implementations, the migration costs can be minimized. In the decentralized model, automatic configuration and ease-of-use features cannot rely on a central element to hide the various technologies behind uniform APIs and middleware. Therefore, service discovery protocols are used to provide a similar way of creating methods for discovering and controlling home devices. Also, in this approach, application protocols and media formats must be agreed upon carefully and special attention must be given to usability and autoconfiguration features.

This decentralized home architecture is based on a layered approach. At the lower level, the home connectivity relies on widely used TCP/IP networking technologies, which are built on top of a variety of link-level bearers and topologies. At the middle level, TCP/IP-based device discovery and control technologies such as UPnP [5] and SLP [9] are used to enable a distributed computing environment. Finally, at the higher level of the architecture are the applications and the media formats that ensure interoperability and usability of the system.

The decentralized approach described above currently appears to have most of the industry momentum. The DLNA and UPnP Forum network architecture follow this approach. Together, these two industry consortiums represent the largest standardization initiatives in the home networking area. Centralized approaches such as the OSGi architecture, although popular in the research community, have still to find widespread support by device manufacturers. Therefore, it seems unlikely that they will form the basis of most home networks in the immediate future.

4.4.2 Local Connectivity

As mentioned earlier, the architecture model presented here for home networks is based on the decentralized architecture model, similar to that chosen in DLNA interoperability guidelines [4]. An example illustration of this architecture model for local network connectivity is depicted in Figure 4.3.

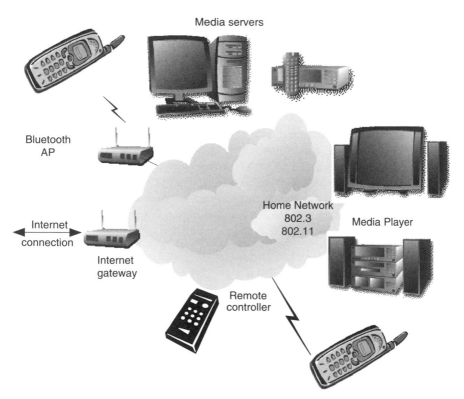

FIGURE 4.3 Overview of local access architecture.

The general decentralized model for local connectivity allows the organization of networks of home devices in two different modes: infrastructure-based and ad hoc (or peer-to-peer) based. In the case of infrastructure-based node organization, network elements such as access points (AP), switches, routers, Dynamic Host Configuration Protocol (DHCP) servers, and so forth, are used to create one or more subnets comprising fixed and mobile home devices, which are all connected with each other through a semi-static network configuration and topology. The infrastructure-based approach requires dedicated nodes to provide network-level services and usually some implicit or explicit network configuration by the users. In the case of ad hoc–based node organization, a subset of home devices connect and communicate with each other opportunistically, whenever user or application actions require it, usually for a short period of time, after which these ad hoc networks dissolve. A common example of this organization mode is point-to-point interactions between two devices, for example when showing pictures taken with a Bluetooth-enabled mobile phone on a Bluetooth-enabled TV set. Both the infrastructure-based and the ad hoc–based modes must present the same network abstraction to higher-level layers (i.e., create an IP layer) to enable applications to operate seamlessly regardless of the link-level network configuration mode. Some link-level technologies enable easier interaction in an infrastructure configuration (e.g., 802.11 [15]), and others provide special features to enable easy ad hoc interactions (e.g., Bluetooth [25], [26]). It is expected that home networks will feature a combination of both the infrastructure and ad hoc modes of node organization.

In the same way that the link-level topologies can vary widely, the link-level bearers may also be very different, as long as they present a network abstraction layer based on IP to higher-layer protocols. Of course, the inherent characteristics of the different link-level technologies have a significant impact on the perceived QoS of the IP connection (e.g., bandwidth, delay, delay variation) by the applications. Therefore, different bearers are suitable to support different types of applications, depending on their QoS requirements. Overall, the issue of QoS support in a home network is a challenging one. Home networks will probably be composed of a multitude of link-level technologies, most of which do not offer QoS guarantees. Therefore, a priority-based QoS mechanism (e.g., [27], [28]) seems more applicable than reservation-based QoS mechanisms in home networks.

Some of the most important link-level technologies for local connectivity in home networks are the following:

- 802.3 (Ethernet 10/100/1000 Mbps) [29]: Ethernet has been the most common link-level technology in traditional IP-based Local Area Networks (LANs). It is a well established technology that offers high data-rates (e.g., Gigabit Ethernet), low delays, and high-reliability. It is more suitable for fixed home devices as it restrains mobility; however, it can be used to connect mobile devices when the users are not moving (e.g., an Ethernet cradle/charger for a digital camera). The most serious issue for its wide adoption in home networks is that it requires cabling the home, which is not always feasible. It is expected, however, that most new home constructions in the future will support high-speed Ethernet cabling in the same way they offer telephone wiring today.

- 802.11 a/b/g (Wi-Fi) [15], [30], [31], [32], [33]: 802.11 is without a doubt the most successful wireless local access technology today. It is one of the big enablers of home networks, as it is an inexpensive technology that can be used to instantly connect home devices scattered around the home, without the need for any costly cabling. It also offers relatively high data-rates (order of tens of Mbps), allows for mobility within the home, and it is suitable for laptop PCs and mobile devices. 802.11 provides seamless support for Ethernet and IP. As with all wireless technologies operating in the unlicensed spectrum, 802.11 is vulnerable to interference from other wireless networks, a problem that is expected to intensify because of the commercial success of this technology. Issues with interference and signal fluctuation may lead to a negative user experience in some applications (e.g., streaming video with strict high-bandwidth, low-delay requirements). Finally, because all wireless technologies operate over a shared medium, appropriate security measures have to be taken to prevent unauthorized use of 802.11-based home networks.

- Bluetooth [25], [26]: Bluetooth is a short-range wireless technology, popular mainly in mobile phones and other mobile devices because it combines low power consumption, low cost, and ad hoc networking features. Bluetooth was created as a cable-replacement technology, and this still remains its main usage today. Bluetooth specifies support for Ethernet encapsulation with the Bluetooth Network Encapsulation Protocol (BNEP) [34] and support for IP as specified in the Personal Area Networking (PAN) [35] profile. Bluetooth offers relatively low data-rates (order of 1–3 Mbps) and is not very suitable for connecting a large number of devices. Although Bluetooth access points can be used to provide local connectivity in the infrastructure mode of node organization mentioned above, it is more likely that it will be mainly used in the ad hoc mode of organization for point-to-point interactions.

- UltraWideBand (UWB; WiMedia) [36], [37]: UWB is an emerging wireless technology for short-range, high-speed communication. It is described as a technology offering high bandwidth (order of hundreds of Mbps) and low power consumption, which makes it a good match for power-constrained mobile devices with rich multimedia capabilities. It is possible that UWB will replace Bluetooth in devices where higher data rates are required. Recently, the Bluetooth SIG announced that it intends to explore the convergence of Bluetooth and UWB technologies. Also, the USB Forum announced the availability of the Wireless USB specification, based on WiMedia UWB technology [38]. The main issues that may delay the wide deployment of UWB in home networks are lack of broad consensus in UWB standardization and regulatory obstacles in some regions outside the United States.

- IEEE 802.15.4 [39]: 802.15.4 is a wireless technology for local access designed to offer low data-rates (order of hundreds of Kbps) and very low power consumption. It offers seamless support for Ethernet and IP and is suitable for connecting a large number of devices. Its characteristics make it a strong candidate for home automation applications involving sensors and actuators, where nodes only need to transmit low data-rate streams but have high requirements for conserving power.

- HomePlug [16]: HomePlug is a wireline technology that supports Ethernet over existing electrical power wires at home. HomePlug adaptors can be plugged in regular electrical power sockets, thus easily creating Ethernet LAN segments connecting all devices scattered around the home. HomePlug technology supports high data-rates (order of tens of Mbps) and can be suitable for transmission of multimedia content. Obviously, the biggest appeal of this technology is that it uses wiring that is available in all current and future homes. Because the power grid often has wiring segments shared by multiple homes, security threats can be considered similar to the case of 802.11. On the negative side, sometimes not all rooms in a home are connected by the same power wiring, which prevents the technology from creating fully connected home networks. Also, often power wiring has poor signal propagation properties and suffers from cross-talk, which may have a negative impact on the perceived user experience.

- IEEE 1394 (FireWire) [40]: FireWire is a serial bus wireline technology, which was originally developed to provide a high-speed serial connection between devices. It supports very high data-rates (order of hundreds of Mbps) and a synchronous mode of operation that can guarantee bandwidth to devices. The technology is suitable for connecting several devices at the same time, and extensions have been proposed to support Ethernet and IP traffic. However, currently it is still mainly used for point-to-point interactions. Also, it is possible that FireWire technology may come under pressure from the ubiquitous and cheaper alternative of USB 2.0, which offers similar data rates.

- Near-Field Communications (NFC) [41]: NFC is a very close proximity wireless technology, which can be used for two-way communication between two devices. Because NFC communication occurs only if the two devices practically "touch" each other, it offers a new, intuitive, and inherently secure user interaction modality. NFC is a technology that grows in importance, and it is expected that it will be incorporated in many mobile devices in the future. NFC itself is suitable for IP traffic exchange, but its significance in home networking can be even bigger as an out-band mechanism used for intuitive initialization of some other in-band bearer, to be used to carry the actual IP traffic.

- Other technologies that may play some role in home networking include USB [38] and HomePNA [42].

Mobile devices can use any of the above wireless or wireline link-level bearers to connect locally with other devices in the home network. Of course, wireless bearers are much more common because they enable users to move freely while using their mobile devices. Compared with fixed home devices connected through wireline bearers, mobile devices using wireless bearers have two important additional problems to overcome: limited power resources and unreliable connections to the home network:

- To reduce power and conserve energy, wireless bearers define low-power modes of operation. In general, the higher the power savings of these modes, the lower the effective throughput and the higher the delay experienced by the mobile devices. In some cases, when a very-low-power mode is entered, the device appears temporarily disconnected from the home network. To optimize energy conservation while maintaining full functionality, often mobile devices must be aided by proxies in the network, which perform certain functionality on their behalf while they are unavailable because they are saving power. Also, it is important that traffic filtering takes place (e.g., at the access point) so that only necessary traffic reaches the mobile devices, to allow them to enter the low-power modes for as much time as possible.

- To deal with frequent disconnections caused by signal fluctuations, interference, and user mobility, it is important that home applications running on mobile devices are designed in a way that tolerates poor network connectivity. In addition, there are proposals for specialized middleware (e.g., [43], [44]) that can provide the necessary end-to-end session support against disconnections and network interface changes (horizontal or vertical mobility) transparently to home applications.

4.4.3 Remote Connectivity

Until recently, the home has been mostly considered as an isolated network environment, where the only outside connectivity has been Internet access from home PCs. Earlier remote access to the home had to be implemented over telephone lines, making it expensive to use and providing only narrow-band connectivity. Currently, mobile terminals such as smart-phones are mostly capable of Internet connectivity, and also a significant portion of homes have broadband Internet connection. This creates a connectivity enabler that provides low-cost access with sufficient data rates for many applications. Although cellular Internet connectivity provides nearly global access to the Internet, its bandwidth and the costs of usage have been a limiting factor for the widespread usage of cellular Internet. However, this is changing as cellular device manufacturers have started to include wireless LAN interfaces as complementary access to the Internet. This allows exploiting WLAN hot spots located in many urban locations and in many homes. The existence of WLAN as an optional bearer is expected to create additional pressure for operators to offer flat-fee plans for cellular Internet access, thus lowering the usage cost.

There are, however, challenges on reaching the home network from a remote location. First, a typical broadband connection uses dynamic IP addresses that in many cases are not even public. If the dynamic address provided by an ISP is accessible through a NAT (Network Address Translation) and is known, then remote terminals may connect to the home network. However, as the home network IP address is assigned dynamically, there needs to be mechanism that allows remote nodes to look it up. Domain names (DNS) can be used for this reason if the home network gateway applies dynamic DNS [45]. Dynamic DNS is a technique that extends the domain name mapping system of the Internet. Usually, a domain name has a fixed relationship with a single IP address, and updates to this mapping are infrequent. The DNS system is hierarchical. Inquiries and management of the DNS names inside of the domain is managed internally for each domain. This is exploited in Dynamic DNS where the domain's DNS server can be dynamically updated and the updates do not need to be propagated to other DNS servers in the hierarchy. When a host is using dynamic DNS, it needs to update its current IP address with a Dynamic DNS server. This server responds to incoming inquiries about this name with the updated IP address. Dynamic DNS does not have a clearly standardized behavior, therefore multiple implementations exist on how the IP addresses are updated at the server. These updates can also be done manually, if the ISP provider allows it.

A second challenge is related to traversing firewalls and NATs. NAT is a technique that allows creating a separate private address space for a private network, allowing virtually unlimited internal nodes to access the Internet through a limited (often only one) set of public IP addresses. In practice, the majority of NATs are using both IP address and TCP/User Datagram Protocol (UDP) port numbers in address translation. Typically, NATs allow only outgoing sessions from the hosts residing in the network. All incoming sessions are discarded. Therefore, in order to reach hosts inside these private networks, typically an outside intermediate node is required. Reaching these private hosts requires that they, in advance, contact a rendezvous server and register their presence. Then this rendezvous server assists clients to establish contact to nodes residing in private networks. There are several protocols (e.g., Simple Traversal of UDP NATs (STUN) [46]) for NAT traversal. Firewalls present similar problems as NATs, although the primary usage for firewalls is to prevent unsolicited traffic from entering a private network. In many cases, outbound traffic can also be restricted. In order to allow incoming connections into a firewall-protected network, there needs to be a controlled way of opening holes for certain incoming connections. A standardized way to do this is through the use of the SOCS protocol [47]. Finally, a good review of the NAT traversal issue and protocols can be found in Ref. 48.

Another challenge is to ensure the security of the remote connection. In an enterprise environment, a typical solution is to deploy a VPN (Virtual Private Network) [49] solution between the remote client and the protected network. There are several solutions based on different protocol layers and varying technologies. Currently, the IETF-defined IPEC protocol [21] is commonly used together with some vendor-specific extensions. VPN solutions basically include authentication and access control between clients and the protected network, as well as encryption and integrity check of the transmitted data. VPN solutions typically present very challenging issues on making configuration and set-up easy enough for nonexpert home users. Still, if direct access to the home network is needed, VPN seems to be the preferred secure solution.

While keeping these challenges in mind, we present some alternatives for building mobile connectivity and remote access to home networks. Figure 4.4 shows a high-level architecture for remote access to the home using a mobile device. The mobile

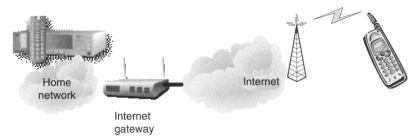

FIGURE 4.4 Overview of remote access architecture.

terminal is connected to the Internet through either a cellular network or through a local WLAN access point (e.g., 802.11b). Today, from the point of view of cost and bandwidth, WLAN access would be preferable for transferring large amounts of content, and cellular access would be more reasonable for transferring single songs or schedule recordings when WLAN is not available. However, the cellular costs are dropping and available bandwidth is increasing, so in the future, full synchronization over cellular networks will provide a feasible choice.

There are several ways to implement actual remote access to the home. One option would be using a third-party service (e.g., [50]) that allows accessing from a remote location home media content that resides in a home PC. In the rest of this section, however, we will focus on two other connectivity options, a proxy solution and a VPN-based solution, which involve only an Internet gateway at the edge of the home network and not some third-party server somewhere else in the Internet.

In the proxy solution, there is a home proxy located at the Internet gateway device. Inside the home, the proxy uses a distributed middleware (e.g., UPnP) to discover and control home devices. At the proxy, this information and potential control actions are converted into some form of HTML pages that can be accessed through standard Web browsers. The proxy is seen as a Web server by the mobile terminal, which accesses the home services from a remote location. All actions that the mobile terminal accesses from the HTML pages are converted into corresponding UPnP actions and vice versa. In this solution, any Web browser can be used for remote access allowing in practice any terminal to be used. Also, authentication and authorization mechanisms are well-known and HTTP traffic is typically able to traverse firewalls. Well-known encryption protocols such as TLS/ Secure Sockets Layer (SSL) [22], [23] can be used to provide adequate level of security. On the other hand, the proxy solution may provide inefficient control of devices and limits the user interface and actions that can be performed remotely only to methods that can be supported by Web browsers.

An alternative solution could be provided based on VPN technologies [49]. VPN creates a protected tunnel between the mobile terminal and the home network using, for example, IPEC [21]. This approach requires a VPN gateway at the Internet gateway and also a VPN client at the remote terminal. In this scenario, the terminal can contact the home devices much like being locally present in the home network, except with increased delay and lower data rates. Although it is possible to use local service discovery protocols over VPN, this may not be the optimum solution for maintaining the status of the network. For instance, UPnP maintains the status of available services by sending advertisements and discovery messages that are multicast to all devices. Although for local networks these periodic and repeated messages may not be a considerable burden, they may cause a significant load for VPN tunnels over remote networks. Therefore, there is a clear need for solutions that allow preventing unnecessary signaling to enter the VPN tunnel but at the same time provide adequate information for the remote terminal to access the home services. The UPnP forum [50] is currently working on this problem domain to provide more appropriate solutions for remote access. The VPN-based solution for remote access requires that a VPN gateway service is available at home, but its configuration is not an easy task for the ordinary consumer. On the other hand, the VPN-based solution allows the terminal to use exactly the same applications and user interfaces that can be used when accessing the home network locally. From the user point of view, this makes remote access of home services easier to use.

Regardless of the chosen approach, usability and security remain vital issues for any remote access products. In both approaches, the actual use of home services can be pretty easy for the consumers, but the initial set-up of the VPN gateway or the home proxy must be made easy first. The main responsibility will remain with the device vendors to create, for example, easy-to-setup wizards; however, innovative ways of transmitting security and network settings from the new devices are required. One alternative would be that ISPs would provide a remote access service as an additional service for their broadband access service.

Remote access to home is an important service in the mobile life that we are living. All approaches presented earlier can coexist, and in the future we shall see more advanced services that will bring additional value for the consumers.

4.5   CONCLUSIONS

In this chapter, we presented an overview of issues related to home networking, with an emphasis on issues related to mobile devices. We presented some basic use cases for mobile devices to illustrate the requirements posed on home networks, both for local and for remote access. We described the challenges facing both the manufacturers and consumers in creating and using such home networks. Because the users of these systems are nonexpert consumers, we believe that the most important challenge is creating technology that makes home networks easy-to-use, self-configuring, and self-healing. Finally, we presented a decentralized architecture and an overview of technologies that can be used to enable local and remote access to home networks using mobile devices. We believe that a decentralized network architecture that builds on popular link-level bearers (e.g., Ethernet and 802.11) and on widely available Internet-based technologies (e.g., TCP/IP, HTTP) has a high potential of success because it allows users to build their home networks gradually while lowering the cost of ownership of such systems as much as possible.

REFERENCES

1. LonMark International. Available at http://www.lonmark.org.

2. X10 Powerline Carrier Technology. Available at http://www.x10.com/support/technology1.htm.

3. R. Lea, S. Gibbs, A. Dara-Abrams, and E. Eytchison, "Networking home entertainment devices with HAVi." *Computer*, Vol. 33, No. 9, Sep. 2000, pp. 35–43.

4. Digital Living Network Alliance (DLNA), "Home Networked Device Interoperability Guidelines v1.0." June 2004.

5. UPnP Forum, "UPnP Device Architecture 1.0.1." December 2003.

6. P. Dobrev. D. Famolari, C. Kurzke, and B. A. Miller, "Device and service discovery in home networks with OSGi." *IEEE Communications Magazine*, Vol. 40, No. 8, pp. 86–92, Aug. 2002.

7. Project Aura home page. Available at http://www-2.cs.cmu.edu/ ~ aura/.

8. H. Schulzrinne, X. Wu, S. Sidiroglou, and S. Berger, "Ubiquitous Computing in Home Networks." *IEEE Communications Magazine*, pp. 128–135, November 2003.

9. E. Guttman et al., "Service Location Protocol," V. 2, IETF RFC 2608, June 1999.

10. J. Rosenberg et al., "SIP: Session Initiation Protocol." IETF RFC 3261, June 2002.

11. P. M. Corcoran, J. Desbonnet, P. Bigioi, and I. Lupu, "Home network infrastructure for handheld/wearable appliances." *IEEE Transactions on Consumer Electronics*, Vol. 48, No. 3, pp. 490–495, Aug. 2002.

12. X. Li and W. Zhang, "The design and implementation of home network system using OSGi compliant middleware." *IEEE Transactions on Consumer Electronics*, Vol. 50, No. 2, May 2004.

13. S. Cheshire and B. Aboba, Dynamic Configuration of IPv4 Link-local Addresses. IETF Internet draft, Zeroconf, March 2001.

14. Droms R., "Dynamic Host Configuration Protocol (DHCP)", IETF RFC 2131, March 1997.

15. ANSI/IEEE 802.11, "802.11std. Wireless LAN Medium Access Control and Physical Layer Specifications." August 1999.

16. HomePlug Powerline Alliance. Available at www.homeplug.org.

17. Wi-Fi Alliance, "Wi-Fi Protected Access (WPA)." October 2002.

18. IEEE 802.1X, "802.1x-2001—Port Based Network Access Control." June 2001.

19. IEEE 802.11i, "802.11 Amendment 6: Medium Access Control Security Enhancements." July 2004.

20. Bluetooth Special Interest Group, "Bluetooth Security Architecture." White paper, version 1.0, 15 July 1999.

21. IETF Network Working Group, "RFC2401: Security Architecture for the Internet Protocol." November 1998.

22. IETF Network Working Group, "RFC2246: The TLS Protocol, v1.0." January 1999.

23. IETF Network Working Group, "RFC2818: HTTP over TLS." May 2000.

24. UPnP Forum, "UPnP Security Ceremonies Design Document v1.0." October 3, 2003.

25. Bluetooth Special Interest Group, "Bluetooth Core." Specification of the Bluetooth System version 1.1, February 2001.

26. Bluetooth Special Interest Group, "Bluetooth Core." Specification of the Bluetooth System version 1.2, November 2003.

27. IEEE 802.1Q, "IEEE standard for local and metropolitan area networks—Common specifications—Virtual Bridged Local Area Networks." May 2003.

28. WMM Specification, Wi-Fi WMM (Wireless Multimedia) Specification. Wi-Fi Alliance, March 2004.

29. IEEE 802.3, "Local and Metropolitan Area Networks—Specific Requirements—Part 3: Carrier Sense Multiple Access with Collision Detection (CSMA/CD), Access Method and Physical Layer Specification." March 8, 2002.

30. Wi-Fi Alliance. Available at www.wi-fi.org.

31. IEEE 802.11a (Supplement to IEEE 802.11, 1999 Edition), "Local and Metropolitan Area Networks—Specific requirements—Part 11: Wireless LAN Medium Access Control (MAC) and Physical Layer (PHY) Specifications: High Speed Physical Layer in the 5 GHz Band." Reaffirmed June 12, 2003.

32. IEEE 802.11b (Supplement to IEEE 802.11, 1999 Edition), "Local and Metropolitan Area Networks—Specific Requirements—Part 11: Wireless LAN Medium Access Control (MAC) and Physical Layer (PHY) Specifications: Higher Speed Physical Layer (PHY) Extension in the 2.4 GHz Band." Reaffirmed June 12, 2003.

33. IEEE 802.11g (Supplement to IEEE Std 802.11, 1999 Edition), "Local and Metropolitan Area Networks—Specific Requirements—Part 11: Wireless LAN Medium Access Control (MAC) and Physical Layer (PHY) Specifications Amendment 4: Further Higher Speed Physical Layer (PHY) Extension in the 2.4 GHz Band." June 27, 2003.

34. Bluetooth Special Interest Group, Bluetooth Network Encapsulation Protocol (BNEP). February 2003.

35. Bluetooth Special Interest Group, Personal Area Networking (PAN) Profile, v1.0. February 2003.

36. Wi-Media Alliance. Available at www.wimedia.org.

37. Porcino D. "Ultra-Wideband Radio Technology: Potential and Challenges Ahead." *IEEE Communications Magazine*, July 2003.

38. USB Forum. Available at www.usb.org.

39. Zigbee Alliance. Available at www.zigbee.org.

40. 1394 Trade Association. Available at www.1394ta.org.

41. NFC Forum. Available at www.nfc-forum.org.

42. HomePNA. Available at www.homepna.org.

43. V. Zandy, B. Miller, "Reliable Network Connections." ACM MOBICOM'02, September 2002.

44. J. Salz, A. Snoeren, H. Balakrishnan, "TESLA: A Transparent, Extensible Session-Layer Architecture for End-to-end Network Services." 4th USENIX Symposium on Internet Technologies and Systems (USITS'03), March 2003.

45. IETF Network Working Group, "RFC 2136 Dynamic Updates in the Domain Name System." April 1997.

46. J. Rosenberg, J. Weinberger, C. Huitema, R. Mahy, "STUN—Simple Traversal of User Datagram Protocol (UDP) Through Network Address Translators (NATs)." IETF RFC3489, March 2003. Available at http://www.ietf.org/rfc/rfc3489.txt.

47. M. Leech, M. Ganis, Y. Lee, R. Kuris, D. Koblas, L. Jones, "SOCKS Protocol Version 5." IETF RFC1928, March 1996. Available at http://www.ietf.org/rfc/rfc1928.txt.

48. B. Ford, P. Srisuresh, and D. Kegel, "Peer-to-Peer Communication Across Network Address Translators." USENIX, April 2005.

49. C. Metz, "The latest in virtual private networks: part I." *IEEE Internet Computing*, Vol. 7, No. 1, pp. 87–91, Jan.–Feb. 2003.

50. ORB networks. Available at http://www.orb.com.

51. UPnP Forum. Available at www.upnp.org.

5

GENERIC ACCESS NETWORK TOWARD FIXED–MOBILE CONVERGENCE

CLAUS LINDHOLT HANSEN

Unlicensed Mobile Access (UMA) is different from other technologies that deliver voice and data over IP/Wi-Fi in that call and revenue control still lies within the operators' core network—you make calls over your Wi-Fi router, but service is still delivered by your GSM operator.

UMA is one of the most compelling technologies for a converged, all-IP network and service evolution because it provides mobile service using a radio technology that requires no license, typically within the industrial, scientific, and medical (ISM) band, such as Bluetooth or IEEE 802.11, while offering a completely *seamless service* to the end user as he moves between the macro Global System for Mobile (GSM) access network and Generic Access Network (GAN). GAN is part of the 3rd Generation Partnership Project (3GPP) R6 standards.

What are the benefits, then, of GAN? Operators will be able to:

- Ability to extend the radio access network without site acquisition, radio planning, licensed spectrum (i.e., at very low investment and operating cost)
- Give good service to end users at places where they live or work (this is shown to be important for customer satisfaction)
- Expand the services offered over fixed broadband with mobility

For end users, the benefits depend upon how the operators wish to offer the service, in terms of pricing, however, some benefits are pretty clear:

- The user has the same services and number as on GSM

Technologies for Home Networking. Edited by Sudhir Dixit and Ramjee Prasad
Copyright © 2008 John Wiley & Sons, Inc.

- Better coverage and voice quality at home or at work
- Seamless transitions (roaming and handover) between GAN and the public GSM access network

Of course, there are challenges with a significant new technology such as UMA/ GAN. The most important is the lack of compatible handsets, initially. The development of the UMA/GAN standard provides the technical basis, and significant interest from the operator community provides the commercial motivation; consequently, it is expected that major handset vendors will bring forth products in small quantities in 2005, with volumes expected from 2006.

5.1 TRENDS IN THE INDUSTRY

UMA has met with tremendous interest from the telecom operator community looking for better and more cost-effective ways to deliver mobile services to users. Many of the worlds leading operators are deploying or trialing UMA systems, and system and handset vendors are bringing compatible products to the market.

In short, UMA allows operators to deliver GSM/General Packet Radio Service (GPRS) services over a fixed broadband network to a dual-mode Wi-Fi handset in a completely seamless manner. The necessary protocols were developed in the UMA forum by leading vendors such as Ericsson and have now been adopted as part of 3GPP R6.

The beginning of the 21st century has seen unprecedented development in two key areas within telecommunication: in *mobile* networks, the exponential growth in mobile users and the deployment of new mobile networks (e.g., 3G); in *fixed* networks, the vast growth of broadband deployment has had major impact on the way in which we enjoy the Internet, teleworking, and other new services.

For all their success, mobile and broadband networks also show drawbacks: When demands rise, it may be difficult for the normal GSM network to provide good enough *coverage* using limited, licensed spectrum. And even as IP broadband capacity grows and delivers Internet, telephony, remote working access, and so forth, it still cannot deliver a call to your *mobile* phone.

5.2 STANDARDIZATION

How to deliver a GSM service over an IP connection? How to spread out GSM coverage into homes and offices without investing in spectrum licenses and expensive and work-intensive base station installations?

These were questions that spurred the UMA Consortium into action. As a result, a set of specifications were released in September 2004. These describe how a standard, seamless GSM/GPRS service may be delivered via IP and unlicensed radio. The user is able to freely roam between the public radio access and, for example, his Wi-Fi network, using the same number and enjoying the same services.

The specifications have since been adopted by 3GPP, the standards body that governs all 2G and 3G mobile specifications, under the acronym GAN (Generic Access Network). This fact indicates that the telecom industry believes that this may be a significant technology in the evolution of modern networks.

5.3 GAN OVERVIEW

This chapter intends to provide the user with an understanding of the GAN system concept.

Figure 5.1 describes how GAN is a third mobile access technology to complement GSM/EDGE Radio Access Network (GERAN) and UTMS Terrestrial Radio Access Network (UTRAN). One of the governing prerequisites from the outset was that the end user, while moving between these three access networks, should

- Remain connected to the same mobile core network
- Enjoy the same services; be reachable on the same number, and so forth
- Experience no break in service when moving between a GAN coverage area and, for example, GERAN

Figure 5.2 shows the network architecture for GAN, and it also shows that the Up interface is the core of the standard, that is, how the terminal communicates with the network, represented by the Generic Access Network Controller (GANC) (the UMA name is UMA Network Controller [UNC]). The Up interface assumes that the handset has a possibility to exchange IP packets with the GANC. How this is achieved is described in other standards, for example, the Bluetooth standard for Personal Area Network (PAN) or the Wi-Fi standard (802.11) and in different broadband access standards (e.g., Cable, xDSL, etc.).

The figure also shows that the GANC uses the standard A and Gb interfaces toward the mobile core network. These are the same interfaces that an ordinary

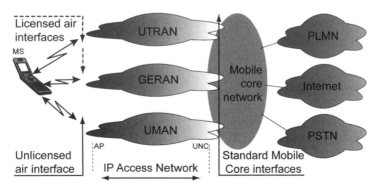

FIGURE 5.1 GAN/UMAN is here seen as a parallel access technology to the well-known GERAN (2G) and UTRAN (3G) networks.

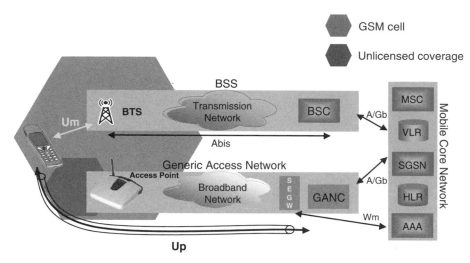

FIGURE 5.2 The GAN architecture. The GAN Controller provides the A and Gb interfaces to the core network and performs the media and protocol gateway functions. The security gateway terminates one end of the IPsec tunnel toward the handset. The access point may be an existing Wi-Fi or Bluetooth device or a GAN-specific box, depending on the operator's preference.

GSM access network uses, so the core network does not need to be aware that the GANC represents another type of access network. It can treat it as if it is an ordinary GSM access network, and all the main GSM services will work, including all packet-based services based on IMS.

Because the Up interface may traverse unsafe IP networks, an IPsec tunnel between the handset and the Security Gateway (SEGW) in the network protects the information sent on the Up interface. All traffic over the Up interface is sent inside this IPsec tunnel and then switched or routed between the SEGW and the GANC. The figure also indicates that GANC–SEGW uses the Wm interface toward an Authentication, Authorization, and Accounting (AAA) server. Only a subset of the Wm interface is used to authenticate that a user trying to connect for service is allowed to do so.

The GAN standard describes how to achieve seamless mobility, handover, and roaming, summarized in the following sections.

5.3.1 Security

GAN security is based on the security mechanisms defined for the 3GPP Interworking WLAN IP Access scenario. The IPsec tunnel protects all control signaling and user plane traffic between the handset and the network in the Up interface. The IPsec tunnel needs to be established before the handset is able to communicate with the GANC, and the handset is authenticated at the tunnel establishment using Subscriber Identity Module [U(SIM), i.e., SIM is 2G or 3G] credentials in a similar way as in GERAN/UTRAN. The protocols used for this part, the Internet Key

Exchange version 2 (IKEv2), Extensible Authentication Protocol/Subscriber Identity Module (EAP/SIM), and Extensible Authentication Protocol/Method for UMTS Authentication and Key Agreement (EAP-AKA), are defined in Internet Engineering Task Force (IETF) specifications.

5.3.2 "Discovery" and "Registration"

The access network between the handset and the GANC is based on IP protocols. This means that the handset in principle has access to different GANC nodes in the GAN. Two procedures called *Discovery* and *Registration* are used to allocate the best possible GANC for the handset in the current location.

Discovery is used between the handset and the Provisioning GANC, which is the initial point-of-contact in the GAN. The handset can either be provisioned with address information about the Provisioning GANC or it can derive the address information from information in the (U)SIM. The main task of the Provisioning GANC is to allow access to the GAN and allocate a Default GANC to each mobile station (MS). This allocation is based on subscription information available in the Provisioning GANC and information provided by the handset during the Discovery procedure. The best Default GANC could be for example the closest GANC to the place the user is living at.

The Default GANC is the main point-of-contact for the handset in the GAN. Whenever the handset tries to access GAN from a new location, it will initiate Registration to the Default GANC and provide information about the current GERAN or UTRAN cell. The Default GANC then decides which GANC can best serve the handset at the current location and the Default GANC may redirect the handset to another Serving GANC. In this case, the handset initiates Registration to that Serving GANC.

The Default GANC can also accept the registration, and in this case it becomes the Serving GANC for the handset in that location. If the right Default GANC has been allocated for the user, then this would happen in most cases. When the registration is accepted, the Serving GANC returns the relevant "GAN System Information" to the handset on the established connection meaning that this information is not broadcast in GAN. The GANC also stores information about the handset for mobile terminating procedures like Paging.

The Serving GANC has normally a connection to the Mobile-services Switching Center (MSC) that controls the macrocell the user resides in. This makes it easier to support handovers between the GAN and the macronetwork if this will become required. So when the mobile requests service from this GANC, it accepts the service request. The mobile then stores the address to this GANC together with the current cell information in a table, so that the next time it connects to the GAN within this macrocell, the request will go directly to this Serving GANC. If this GANC reports a different Location Area than the macronetwork, the mobile will send a location update message via the GAN, and when all this is done, the mobile shows the symbol that it now is served by the GAN. As long as this symbol is shown, all originating and terminating traffic will be routed via the GAN.

5.3.3   Rove In and Rove Out

Roving is the term used in the GAN standard for roaming between the WLAN cover-age and GERAN/UTRAN. *Rove in* means that the handset starts communicating actively using the protocols in the Up interface and these protocols start serving the upper layers in the Handset. *Rove out* means that the handset stops communicating using the protocols in the Up interface and the relevant GERAN/UTRAN protocols are used instead and serve also the upper layers in the Handset.

5.3.4   Transparent Access to Services in the Mobile Core Network

The Up interface protocols provide transparent support for services in the mobile core network. This is achieved by tunneling all the upper-layer messages, like mobi-lity management, Short Message Service (SMS), Call Control, and Supplementary Services, over the Up interface and then interworking these to the existing mechan-isms in the A and Gb interfaces.

5.3.5   GPRS Support in GAN

The Up interface also supports transport of GPRS control signaling and user plane traffic using specific procedures. The procedures and design principles for GPRS support in GAN allow the network to support a very large number of handsets, as there is no need to keep handset-specific data in the GPRS part of GANC for idle handsets. Furthermore, flexible load distribution in the network is also supported.

5.3.6   Location Services

The location of a handset registered to the GAN can be found out multiple ways that are based on information received from the handset during registration and infor-mation configured in the GAN and accessible from the GANC. The handset indicates to the GANC the current camped GERAN or UTRAN cell, and already this gives some guidance about the location of the handset. The handset also includes the MAC address of the current WLAN AP being used, and the GANC can use an exter-nal database to map these to an exact geographic location of the AP (as the handset is close to the AP because of the short-range radio). The handset may also be able to report the geographic location to the GANC. Furthermore, the SEGW sees the public IP address used by the handset, and in some scenarios, this information can be used to find out the geographic location of the Handset. The handset may also include street address information in the registration request to the GANC, and this can be mapped to a geographic location using external databases.

5.3.7   Emergency Services

In GAN, the operator has the possibility to indicate to the mobile stations whether emergency calls shall be placed over GERAN/UTRAN or over GAN. If GERAN/

UTRAN is indicated by the GANC and GERAN/UTRAN coverage is available at the handset location, emergency calls are placed over the GERAN/UTRAN network and the existing location determination services are used. In all other cases, the emergency calls are placed over the GAN and location determination mechanisms are used to guide the core network in routing to the right Public Safety Answering Point (PSAP) and also to deliver more exact location information to the mobile core network when requested.

5.3.8 GAN Protocol Architecture

The protocol architecture for the GAN Circuit Switched (CS) domain control plane is shown as an example in Figure 5.3. The new protocols defined in the GAN standard, Generic Access Circuit Switched Resources (GA-CSR), and below serve mobility management (MM) and layers above and have replaced the relevant GERAN and UTRAN radio resource management protocols.

It can also be seen that the new protocols for CS control plane are transported using a Transmission Control Protocol (TCP) connection between the handset and the GANC using the security provided by the IPsec tunnel between the handset and the GANC–SEGW.

It can also be seen that all protocol layers above GA-CSR, that is, Mobility Management (MM), Call Control (CC), Supplementary Services (SS), and SMS, are unmodified and are transported transparently between the handset and the Mobile Switching Center (MSC). These protocols are tunneled in the GA-CSR protocol in the Up interface and transported over the A-interface using standard mechanisms.

The A interface is the interface between the MSC and the GANC. The signaling over the A-interface is done according to the Base Station System Application Part (BSSAP) protocol. BSSAP uses the Message Transfer Part (MTP) and the Signaling Connection Control Part (SCCP). The BSSAP messages can be divided into two

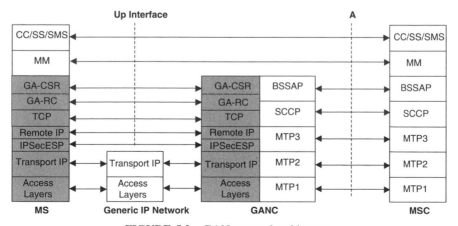

FIGURE 5.3   GAN protocol architecture.

categories: transparent DTAP (Direct Transfer Application Part) messages sent to the handset and nontransparent BSSMAP (Base Station Subsystem Management Application Part) messages sent to the GANC. The GANC performs the needed interworking between the BSSMAP and the GA-CSR protocols.

5.3.9   Bluetooth or Wi-Fi?

Around year 2000 when some operators and vendors had initial discussions on what later became GAN, Bluetooth was already finding its way into many mainstream handset models, and it was therefore thought to be the ideal technology for this application. A year or two later saw the great boom in deployment of IEEE 802.11 access points for wireless PC-to-LAN connections in homes and offices, and there ensued a great debate in the industry on which of these technologies would survive. In the UMA forum, and later in the GAN work in 3GPP, the stance was taken that there should be no dependency on the physical radio layer. This remains an issue between the handset and the access point; Up itself is agnostic to this.

At the time of writing, most trials and early deployments take place with Bluetooth; however there seems to be consensus that Wi-Fi will prevail, mainly due to the huge number of access points, routers, modems, and so forth, that are already installed in homes, offices, and hot spots—a "ready-made" infrastructure on which to offer a GAN-based service. (There has not been installed any significant number of Bluetooth access points, due to lack of viable applications for these.)

Particularly with Wi-Fi equipped handsets, operators who subsidize these are interested in controlling what they are used for. Operators typically see no interest in a VoIP client on the handset being able to provide a voice service, bypassing the operator's core network and, more important, billing system. For these operators, GAN is an opportunity to offer a voice service over local WLAN networks with good coverage while ensuring that the traffic remains controlled in their core networks.

5.4   BENEFITS WITH THE GAN TECHNOLOGY

Let us first recount a number of important aspects of GAN:

- No radio license is required, because GAN operates in the license-free spectrum. (In fact, GAN will allow nonradio devices to connect to the GSM core network as long as the Up specification is adhered to.)
- No radio planning is required. Current license-free technologies such as Bluetooth and IEEE 802.11 are "self-planning."
- Range is limited compared with the macrotechnologies of GERAN and UTRAN. Typically, one can expect 20–30 m indoors and up to 100 m outdoors or larger indoor spaces.

- Instead of base stations, GAN utilizes WLAN access points that are either already in place or set up for the purpose. These are typically around the 50–100 USD price point, or even lower.
- GAN uses IP as its bearer service; that is, it may be delivered over DSL, cable, Ethernet, or other networks already in place to homes and offices.

Which of the benefits listed here are relevant to which actors depends to a large degree upon which business models are in place, which customer segments are in focus, and so forth.

5.4.1 Operators

In many markets, it has been found that mobile subscribers are relatively unfaithful; and one of the main parameters to retain customers is good coverage where subscribers live and work.

The GAN technology enables the operator to significantly improve indoor coverage, especially in North America as GSM in this area mainly operates on the 1900 MHz spectrum with limited reach in large residential areas. By deploying GAN, the operator can deploy and extend the indoor coverage locally without impacting the end user behavior or functionality while also providing a seamless user experience between the networks. Moreover, he can do this in a very cost-efficient fashion due to the low cost of unlicensed access point technology.

Another opportunity for the operator is to deploy mobile voice services over the broadband networks that are now becoming broadly deployed. There are several alternatives of how to deploy voice services over broadband access, none of these, however, offer the service richness of the mobile network, let alone the capability for seamless transition between broadband and mobile network. GAN combines the use of the broadband network with the mobile core network infrastructure, and the mobile handset remains the prime device. The existing functionality, such as charging, authentication, and end user administration in the mobile core network, is reused, while impact is limited to configuration of the new access network.

With GAN, a converged end-user offering can be constructed that both leverages mobile telephony and broadband.

5.4.2 End User

Directly, operators stand to gain most with the GAN technology, but if typical market mechanisms prevail, some of these benefits will be passed on to the end user. We speculate (and initial service offerings based on GAN confirm) that end users will be offered a better quality mobile service at home at a lower price.

With GAN, the end user experience remains the same in the WLAN domain as in the wide area radio domain. The end user has *one* phone with *one* number that works independently of access method and location. The GAN-enabled handset is preconfigured by the operator and does not require any configuration by the end user more than an ordinary GSM handset. The end user experiences functionality transparency (same services independently of network) and seamless mobility between the two domains with roaming and handover in both directions.

5.4.3 Terminal Availability

Ultimately, the success of GAN hinges on the availability of compatible handsets. Standardization is therefore a key element while introducing new protocols into the handsets as this guarantees interoperability between the handsets and the infrastructure.

Terminal and network vendors have taken an active part in the 3GPP GAN standardization process. The standard minimizes the impact on the terminal and leverages on already existing implementations in the handsets.

During early 2006, we saw several tier 1 and tier 2 handset vendors announcing products for the commercial market. This will remove one of the greatest uncertainties around the whole GAN technology.

5.5 PRACTICAL EXPERIENCES

A number of operators have conducted UMA/GAN network trials, as well as some trials with pre-UMA solutions, with limited numbers of participants. Typical findings are

- Setting up the service, plugging in the access point, registering the handset, and so forth, are usually straightforward, a "plug and play" experience.
- Coverage, even with early pre-UMA handsets with low-power Bluetooth radio, is sufficient for residential uses and comparable with normal cordless phones.
- Service and user experience is indistinguishable from what is found via the macronetwork, except for battery life, which with some early implementations was significantly lower with UMA enabled.

Most trials have been concerned with user experience as this is important to create a viable service offering. Because GAN connects into the normal GSM core network and makes use of standard IP connectivity, there has not been much trial activity on that side. These are typically areas that are covered in interoperability testing (IOT) prior to commercial launch.

5.6 IMPACT ON NETWORKS AND PROCESSES

Although GAN makes use of the standardized A and Gb interfaces to the mobile core network, already serving a billion users today, there are certain things to take into account.

Even though the actual core network, hardware and software, does not need changing to work with GAN, operators will probably need to change certain *configuration data* in order to facilitate, for example, handover between GAN and GERAN.

Also, the key components of the GAN itself need to be installed (e.g., the GANC, the security gateway, etc).

Thus, there will be some impact on the operator's planning and operational processes. Because a GANC is largely going to appear as "just another BSC," this impact is probably going to be modest compared with the work involved in launching a completely new type of service. This will involve many customer-oriented processes as well as back-office areas such as billing, and even regulatory areas such as location capability for emergency calls.

The IP network will largely be unaware that a mobile service is traveling across it by way of Up protocol. Despite early caution, it seems that GAN service works quite well over a "best-effort" IP connection as offered by residential broadband connections. However, at the time of writing, data from practical deployment in large scale is not available.

5.7 DISCUSSION

It is always interesting, when a new network concept arrives, to try to predict which impact it will have in the technical arena, in the business world—and of course for end users.

Now when several vendors announce UMA/GAN compatible handsets, we believe that GAN presents the ideal way to offer voice and data connectivity, due to the seamless user experience, which follows from being connected to the same core network regardless of access type. In this aspect, other "voice over WLAN" protocols do not deliver the same transparency, as they rely on the user being served by a different core network when in "WLAN-mode."

GAN is a complement to current mobile access methods, GERAN and UTRAN, and because it utilizes all call handling, mobility management, and so forth, from these, it is relatively straightforward to adopt into handsets.

On the business side, GAN offers many compelling features: It allows broadband operators to combat "fixed-to-mobile substitution" by offering a full-spec mobile service to their customer base. It allows mobile operators to (cheaply) grow access capacity more aggressively than possible with GERAN and UTRAN alone. As a result, we may see changes in business models and marketing strategies of various telecom operators.

End users seem to be interested in two things: "What do I get?" and "How much do I have to pay?" Early information seems to confirm that operators with GAN can answer "More" and "Less." Better call quality, more minutes, for less monthly spending. When comparing all the new possibilities open to end users offered by various peer-to-peer solutions, it seems likely that users will choose to use the device they know best (the mobile phone) and they prefer to use it the same way as usual (i.e., connected to the GSM core network).

5.8 EVOLUTION OF GAN

Fixed–Mobile Convergence (FMC) has been a topic debated with varying intensity during the latest decade. FMC is one of the telecom terms with the widest scope of interpretations. One bill for both fixed and mobile calls is FMC. Redirecting calls between GSM and Plain Old Telephony Service (POTS) is FMC. More "deep" convergence, however, is offered by technologies such as IMS and UMA.

IMS is a core network evolution that will offer current voice and data services, as well as new combinations of these, through different access networks.

Both UMA and IMS belong to the domain of fixed–mobile convergence, and the question is therefore often put: How can UMA evolve into IMS?

IMS (IP Multimedia Subsystem) is an architecture currently being standardized mainly in 3GPP but also partly in the European Telecommunications Standards Institute (ETSI)/Telecoms & Internet Converged Services & Protocols for Advanced Networks (TISPAN). The overlying idea is to evolve a common core network architecture for fixed and mobile access, voice and data services, and, ultimately, a framework for developing services that today are only available on the Internet or not at all. As an example, IMS can allow users during a mobile voice call to take pictures or video clips and share them in real time.

IMS is an evolution of the core and service layers, whereas UMA is an evolution of the mobile access network. Between these layers there exist standard interfaces, such as Gb or Iu, which to a large degree prevent evolution in one layer from impacting the other layer.

Because UMA and IMS belong to different layers of the network connected by standardized interfaces, there is virtually no dependency either way. It is quite feasible to deploy a modern IMS core and make use of current access technologies; likewise, it is possible to reap all the benefits of a UMA access network without committing to upgrade the core to IMS. In our view, this is an ideal situation because it offers operators flexibility in terms of how they wish to evolve access and core networks and does not tie them down to a set path once they decide to deploy UMA.

Evolution to 3G is another topic often discussed in relation to GAN. GAN is specified by 3GPP as connecting to a "2G" core using A interface, so when will it be 3G compatible? There are early studies in 3GPP in this area, however, in our view GAN is already compatible with 3G core networks and handsets: Our belief is that virtually all modern 3G core networks need to provide A and Gb interfaces in order to bring in traffic from existing 2G access networks; the GAN can thus be connected. It is furthermore our view that virtually all 3G handsets will be 2G (and therefore GAN) compatible as well.

5.9 CONCLUSIONS

GAN is backed by a number of dominant handset and system providers on the market; obviously driven by the great interest that has been demonstrated by operators. It is not surprising that a technology with clear benefits for both fixed and mobile operators (and not forgetting end users) is met with keen interest, and now that this technology is proved to work and the protocols are standardized, there are no technical barriers for success.

Therefore, we argue that GAN is likely to become a dominant technology for delivering mobile service over unlicensed radio.

6

SECURE WIRELESS PERSONAL NETWORKS: HOME EXTENDED TO ANYWHERE

JOHN FARSEROTU AND JUHA SAARNIO

Secure *wireless personal networks* constitute the new communications landscape for everyone. As life goes mobile, we will increasingly rely on wireless connectivity, which is currently evolving toward a rapid, reliable, efficient, cost-effective, and secure means of exchanging information. At the same time, global wireless reachability has become a part of our daily lives, making communicating simple.

The next step is to be able to hook up all our possessions to a seamless, global wireless infrastructure. This will enable us to combine the knowledge we have with the information stored in and acquired by our belongings without geographic limitation. In this way, we increase our reach and awareness through telepresence, enhancing personal security and convenience. Our newfound telepresence may be exploited to best advantage by allowing our personal network (at home, business, or on the move) to extend to include our friends' personal networks; for example, to enable efficient sharing of resources and group communications. The ability to extend our personal networks also has an impact on public safety and citizens cooperating with public authorities when the need arises. Of course, the utility, as a whole, stems from the ability to easily and securely control and shape our personal communications environment. We all need to remain the master of our private property and personal information, even if it is virtually extended beyond local physical constraints to new and remote reaches.

Today, a quiet revolution is taking place in personal communications and networking. Over the next few years, secure wireless personal networks (PNs) will begin to

Technologies for Home Networking. Edited by Sudhir Dixit and Ramjee Prasad
Copyright © 2008 John Wiley & Sons, Inc.

You

And

I

Our Secure
Wireless PN

FIGURE 6.1   Secure wireless personal networks for nomadic users.

emerge that allow us to connect with our personal belongings, family, friends, and their belongings, wherever they may be (Fig. 6.1). Such PNs will enable nomadic users and their devices to connect to remote users and devices. Applications running on local user devices can be extended into end-to-end personal services between users, devices, and other trusted parties. Services may range from low-data-rate health monitoring to high-data-rate file transfer to entertainment to subscription-based multimedia downloads, whether in the home environment or outside of it. In the process, a vast amount of personal information must be transmitted over the Internet and wireless access. This information may be anyone's; yours, mine, a friend's, or the family members'. It must be possible to readily exchange and share sensitive private information, and at the same time we must be able to trust that it remains private and that access is limited to trusted parties (i.e., parties that we have chosen to trust).

To put the problem into perspective, consider that the information may include anything from routine calls, e-mails and Web access to information about personal subscription services and preferences, credit information, health data, multimedia content, even possibly remote dosage monitoring and control for implantable devices. If a person's mobile device is exposed to a virus from the network, does that mean they can no longer trust their health monitoring devices and service? We don't want to expose ourselves to totally new kinds of potential physical attacks due to migrating to using PNs as an integral part of our daily communications.

It is difficult to overestimate the volume of information that may be exchanged via PNs. Many personal applications and services may be low data rate and actual information transfer may be small. For this reason, one might initially expect the traffic volume to be small as well. Indeed, it is probably not interesting for a service provider to offer a personalized service to a single user. It is the collectivity of users that count. The aggregation of information from a vast array of personal devices and sensors, wherever they may be, is potentially enormous. We may soon find ourselves swimming in a sea of radio frequency (RF) from wireless devices conveying all types of information. Importantly, it is a sea that is becoming friendlier and more meaningful.

In this respect, the automated exchange of low-data-rate monitoring and control information between always-on, always-connected devices parallels the emergence of short message service (SMS), which today constitutes as much as 40% to 50% of the traffic volume. The potential with PNs and services though is much more, even before we consider higher data rates. Aggregation of traffic is expected to be an important part of service provision. This will be the job of service centers, whether with existing operators and service providers or start-up secure personal service providers and service centers.

Before practical PNs and services are possible, though, various technical advances are required. This is an area of intense research currently being investigated in the European Commission IST 6th Framework Integrated Project "My personal Adaptive Global NET Beyond" (MAGNET) Integrated Project (IP), the second phase of the IST MAGNET IP [1]. In addition, its mission is to carry the MAGNET vision of personal services over personal networks a step closer to reality through the introduction of pilot services. The development of person-centric applications and services are also an integral part of the European eMobility Technology Platform [2]. The eMobility platform envisions a new paradigm: "The improvement of the individual's quality of life, achieved through the availability of an environment for instant provision and access to meaningful, multi-sensory information and content."

Thus, the increased awareness afforded by telepresence may be combined with sensory input and context awareness to further enhance security and convenience. The success of future personalized wireless services requires more than breaking down technological barriers. Importantly, it also requires building of user trust. This in turn requires transparent, trustworthy, and easy-to-use security mechanisms from top-to-bottom and end-to-end. In this chapter, we extend the notion of home network to a virtual personal (or home) network extending to remotely located devices, thus breaking down the barrier of distance. In short, we could call such a network a "virtual home network."

6.1 A VISION OF A PERSONAL NETWORK

Figure 6.2 provides an overview of the vision for PNs. The vision calls for secure PNs to support the users' professional and private activities, without being obtrusive and while safeguarding their privacy and security.

In many respects, a PN may be considered a PAN (personal area network) or a home network (which is private and secured), but without the geographic limitations [3–5]. PNs may extend from users' personal PANs (P-PANs) to homes and personal vehicular clusters. A PN can operate on top of any number of networks that exist for subscriber services or are composed in an ad hoc manner for this particular purpose. For the purposes of the discussion in this chapter, a distinction is made between applications and services. Specifically, an application may run locally or remotely, but the service is end-to-end. A PN may or may not be needed to run an application, but it is always needed to support personal services.

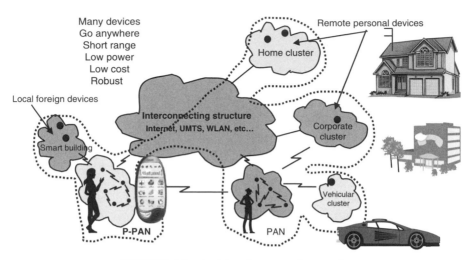

FIGURE 6.2   A vision for personal networks.

6.2 SOME EXAMPLE SCENARIOS

We are already witnessing increasing amounts of data traffic. When personal content creation and sharing is fully enabled, a data tornado will ensue. This will drive up the need for new communication infrastructures and catalyze the creation of new applications and services enabled by the developments brought forth. For the purposes of the discussion in this chapter, we present three example scenarios encompassing three broad mass application domains: (1) health, (2) home, and daily life, and (3) distributed work. We assume that once a trusted personal network has been established, all possible applications can be enabled as though they were running in a secured local environment; therefore, we purposely do not include the specific applications in the remainder of this section.

6.2.1 Health

The health sector includes monitoring of patients anywhere, anytime. Figure 6.3 provides an illustration. Patients may be at home, in the hospital, or on the move, but they are connected to a cooperating network with doctors, nurses, and their friends. The health care scenario extends to emergency situations, patients en route in an ambulance, as well as to emergency and medical staff on the move. Persons in this scenario may be "well" patients just as well as "sick" patients, for example, activity monitoring for sports and training.

Diabetes provides an instructive example of a specific health care scenario. Diabetes is potentially one of the most important health care markets for PNs. By some estimates, it represents as much as 20% of the health care expenditures in Europe. Many people who would otherwise live full and long lives needlessly suffer and die years earlier than they should with treatments available today.

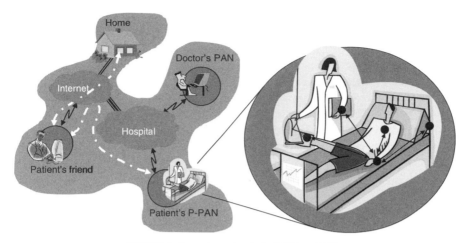

FIGURE 6.3   A patient's P-PAN and PNs.

A significant part of the problem is that it is neither easy nor convenient to continuously and accurately monitor glucose levels. People work, travel, and shop. They may forget or postpone a reading. The readings themselves may vary more than a person realizes. The consequences are significant and cumulative.

Within MAGNET [1], a diabetes scenario has been studied and a multisensor approach is considered for noninvasive monitoring of patients. Ideally, such personal sensor devices are small enough and of low-enough cost to be wearable and even throwaway. Some typical data exchange requirements are summarized in Table 6.1 for sensors including:

- Blood pressure
- ECG
- Respiration
- Thermometer

TABLE 6.1   Data Exchange Requirements for a Diabetes Scenario

Device No.	Type	Data Block Size	Block Rate	Information Rate
0	Co-coordinator (receiver)	Arbitrary	Arbitrary	Arbitrary
1	Blood pressure sensor	64 bit	2 blocks/min	128 bits/min
2	ECG sensor	1024 bit	10 blocks/s	10240 bits/s
3	Respiratory sensor	64 bit	10 blocks/s	640 bits/s
4	Clinical thermometer	8 bit	1 block/s	8 bits/s
5	Pulse ear sensor	8 bit	1 block/s	8 bits/s

Based on these figures, it can be seen that diabetes requires low-data-rate (LDR) communication. Indeed, it can be seen that the total information transfer requirements for the multisensor diabetes application should be no more than about 10–20 kbps, depending on the combination of sensors. Actual data rates would be somewhat higher given the protocol overhead.

The implications of this in terms of networking, security, power consumption, PAN-optimized air interfaces and devices are quite significant. Similarly, for the home and daily life and distributed work scenarios, both high- and low-data-rate information exchange requirements apply.

6.2.2   Home and Daily Life

The home and daily life scenario and markets include the office and home environment. It spans smart shopping, personal convenience, remote interaction with family and friends, entertainment, and security. Applications may range from low-data-rate remote control to high-speed data transfers (e.g., music and DVDs).

The use of nomadic PANs is not limited to homes in this context; it may readily be extended to vehicles and other homes away from home, thus extending the definition of home so long as the key attributes of home environment (e.g., privacy, security) are retained. In this sense, a nomadic PAN may be considered as all of the user devices that operate within a short distance defining a small space about the user that travels with him or her. This portable, private PAN enables secure end-to-end connectivity and services anytime, anywhere, as if it was an extension of the user's home.

There is a lot of efficiency to be gained from collecting context data and linking this to personal data available through networks. One could, for example, foresee a user scenario, where all the data one's personal trusted device sees and all events the device partakes in are recorded as a digital timeline of a person's experiences. This timeline could then be used for a multitude of valuable services improving the person's life. Being able to scroll back to a specific past event when and where needed can prove to be invaluable. Also being able to search for items in a human way by using search strings like "I would like to call that Asian person I chatted with for a long time at the function dinner last Wednesday or Thursday, but what was his name again?" can sometimes make all the difference in the world.

Adding presence information to the digital life recorded and categorizing certain context groupings can also yield interesting efficiency gains. If, for example, a person walks into a shop, and the device can somehow detect this, a precreated home needs list could be automatically loaded and presented to the owner of the device. This kind of automatic needs sensing based on presence and context can be hugely valuable to a user. It also needs to allow personalization to suit one's needs, and learning capability can add to the usefulness over time.

6.2.3   Distributed Work

Distributed work is typified by cooperation between multiple individuals working toward a common goal. These individuals may be scattered over widely different geographic areas and may use various different applications, networks, and infrastructures to exchange views, data, images, and drawings as they work together toward a common goal. Examples of distributed work scenarios include:

- Journalists and mass media covering a story
- Students working on a common project
- Researchers in cooperating projects

PNs enable parties to share resources in everyday life at home or work. This may be important when it comes to impromptu collaboration between, for example, police, fire, and rescue workers. By allowing trusted parties to share resources by coupling two or more personal networks will speed up results. For instance, public servants, such as the police, the fire brigade, and the medical response team, federating their organizational PNs in a crisis situation will lead to heightened control of the situation and efficient use of resources.

Collaborations may also extend beyond personal networks to what may be referred to as federated networks. In federated networks, PNs are extended to remote, mobile, collaborative networks. In this respect, we may consider that our PN has been extended from you and me to our colleagues. The one-to-one connectivity has become a many-to-many virtual relationship. This of course only heightens the need for pervasive security and transferable trust.

6.3 SYSTEM AND REQUIREMENTS

Some devices may be wearable, even throwaway, whereas others, such as handsets or personal trusted devices, are more capable in terms of processing and autonomy. In general though, personal devices, such as those within the P-PAN, need to be trusted, small, highly portable, low cost, and low power, but they may connect and communicate with other nonportable devices and apparatus in their environment. Data rates depend on the application and may vary from as little as a few kbps to perhaps hundreds or thousands of Mbps.

The functional requirements for wireless communication in the P-PAN may be summarized as follows.

- Short range
- Low power
- Low cost
- Good coexistence
- Robust, simple-to-use, and reliable devices
- Small, highly portable, and efficient
- Go-anywhere operation
- Secure and trusted

Communication within the P-PAN [1, 3] is limited to a range of not more than about 5–10 m and, in many cases, 1–2 m may be sufficient. The autonomy of a device may be measured in days to weeks. On the other hand, the goal for simple personal sensor devices may be multiple years. If the power consumption is low enough, energy scavenging (e.g., no batteries) may be possible, which would be ideal for many throwaway devices. Some consider the threshold for energy scavenging to be not more than about 10–20 μW. This is a tough but not an unrealistic objective.

To put it into perspective, let us consider a simple example for a health monitoring application similar to the blood pressure sensor from the diabetes scenario in Table 6.1. In this example, Bluetooth is assumed for short-range wireless communication. The transmission rate (R) is 721 kbps and the power consumption (P) during transmission is of the order 100 mW. For health monitoring applications, operation 24 hours per day, 365 days per year may be required. Transmission of short, 32-byte messages are considered in this example (i.e., sensor data and protocol overhead).

From Table 6.1, sensor updates are transmitted twice per minute. The transmit on-time (T_{on}) was evaluated and found to be about 373 seconds per year: $T_{on} =$ (1,051,200 messages/year $\times$ 256 bits/message)/(721,000 bits/s). The fractional on time (f_{on}) per year is $f_{on} = $ (373 seconds/year)/(31,536,000 seconds/year) $=$ 1.18×10^{-5}. The average power consumption (P_{ave}) may then be evaluated as: $P_{ave} = f_{on} \times P \approx 1$ mW.

Now we look at this example from the perspective of autonomy. Considering a 1.5-V battery, we first calculate the average current (i_{ave}) using the relation $P = iv$ ($i_{ave} = 100$ mW/1.5 V $= 66.7$ mA). This translates into about 6.9 mAh per year (66.7 mA $\times$ 373 seconds per year/3600 seconds/hour $= 6.9$ mAh). Based on this example, it can be seen that multiple years of autonomy could be possible with a small watch cell battery (e.g., 8 mm $\times$ 1.65 mm, 22 mAh). However, for other applications in the same diabetes scenario, such as the respiratory sensor, a reduction of power consumption by a factor of the order 10–100 would be needed to achieve even a couple of weeks autonomy, with the same watch cell battery. Even more would be required in the case of the ECG. Furthermore, we must also consider the protocol and the sensor itself, which may consume more than the radio in some cases. These are challenges for personal sensor devices.

Whereas the cost of more capable devices, such as handsets and PDAs, may exceed 100 euros, the target cost of simple personal sensor devices is nominally less than 1 euro; perhaps as little as a few cents. At the same time, health care applications impose requirements for robustness and reliability even on the simplest of devices. Despite a potential "sea of RF," certain critical information must get through in a timely manner, and this must be the case anywhere, anytime. Good RF coexistence characteristics and use of available spectrum are a must for the future. Otherwise, the very success of wireless systems may become a barrier to continued growth. At the same time, communication from even simple devices must be reliable and secure.

Today, current technology is not adequate to meet the full range of requirements. Devices are too large, costly, and power consuming. Networking is inadequate to support secure PNs in a practical manner. Advances are needed in networking, communication, security, and platforms. Key issues include the ability to provide secure end-to-end wireless personal networks to even simple, low-complexity personal devices, which, although they are simple and inexpensive, may handle some of the most sensitive and highly personal information.

6.4 USER REQUIREMENTS AND SCENARIOS

PNs encompass potentially all of a person's devices being networking enabled and capable of connecting to a network through physical or wireless interfaces. The ultimate vision is that future users will be supported in their private and professional activities by their PNs. It is considered that the PN concept will only become widely accepted when a "sufficient" level of security is guaranteed and maintained. Therefore, high-level security requirements for the PN infrastructure must be defined for both entities: individual (unknown) ad hoc PN user and an existing PAN or PN (or group of them) configuration capable of providing connection to any user type requesting services. Securing PNs is most often seen as a purely technical matter. Better access control policy models, formal proofs of cryptographic protocols, approved firewalls, and so forth, will solve the problems.

The argument presented in this chapter explains that the problem is more complicated involving both technical (see above) and socioeconomic aspects as network externalities, asymmetric information, and so forth, and other dimensions including an infrastructure/social dimension and a private dimension.

6.5 NETWORK ARCHITECTURE

At the heart of secure PNs are the person and their network [1, 5, 6–10]. Devices may be widely distributed and connectivity between nomadic users is ad hoc in nature [6]. Research is ongoing to answer the questions concerning the nature of a PN. What is a PN? How does it differ from PAN? How does it differ from Wireless Personal Area Network (WPAN) legacy networks and what is the state-of-the-art?

Key issues with respect to secure PNs include integration into heterogeneous networks, addressing, PN protocols (e.g., secure remote service discovery), mobility, and handover. The capabilities and limitations of IPv4 and IPv6 for PNs are being studied, and the adequacy of existing protocols (e.g., service discovery) are examined with respect to PNs.

Additionally, the complementary concept of federated networks in relation to PNs is very interesting. Other issues for study include the potential relation to and interaction between PNs and other systems, such as future wireless sensor networks (WSNs), home area networks (HANs), and Wireless Local Area Networks (WLANs) in offices. The reader is referred to the IST MAGNET project [1] for answers to the questions posed above.

6.6   ACCESS AND ACCESS CONTROL TECHNIQUES

The requirements for realization of secure personal networks extend to all layers. This includes the Air Interace (AI) and corresponding choices for the Physical (PHY) and Medium Access Control (MAC) layers. No one solution fits all applications. A flexible approach is considered taking into account PAN optimized Physical and Medium Access Control (PHY-MAC) layer solutions and comparing them with emerging and legacy systems [11–14].

The AI solutions can be divided into two broad categories: low-complexity PHY-MAC solutions for simple, low-data-rate applications, and high-performance solutions for high-data-rate (HDR) applications running on more capable personal devices. Frequency Modulation UWB (FM-UWB) has been selected for low-complexity LDR applications, and Multicarrier Spread Spectrum (MC-SS) has been selected for HDR applications.

MC-SS offers HDR communication with high spectral efficiency and robustness. FM-UWB offers the potential for low-power, low-cost, yet robust communication in the presence of narrowband interference and multipath fading. In this respect, it may be viewed as a form of analogue spread spectrum communication, which also offers the potential for enhanced security and privacy of wireless links in the P-PAN. Enhanced robustness and privacy may be especially important for supporting personal services where reliability is at a premium, such as health monitoring and emergency services.

6.7   SECURITY

Security for LDR devices (e.g., sensors) typically consists of five components: sensing hardware, memory, battery, embedded processor, and transceiver [11, 15–17]. These components affect the performance of the sensor and ultimately that of the network. For low-complexity personal sensor devices, the goal is to design an adaptive security protocol and crypto algorithm that provides authentication and confidentiality under the constraints of code size, CPU, and memory size.

A flexible, adaptive approach is required to provide security in LDR environments based on service-aware adaptive security architecture, capable of providing three different levels of security that best match with the actual needs of LDR services, without overloading the constrained LDR devices. Within the P-PAN, the processing load is handled by more devices, such as handsets and PDAs, capable of supporting mobile gateway and personal firewall functionalities.

6.8   DEVICES AND SERVICE PLATFORMS

A vision of PANs should include multiple personal devices with various capabilities in terms of data rates and functionalities [1]. Some devices may support only one PAN air interface (low, medium, or high data rate), and some others may need more than one air interface based on their application requirements. More capable gateway devices would support the PAN air interfaces along with one or more legacy technologies to connect to the core network. Such devices play a potentially key role in terms of secure PNs.

Today, security for personal devices and platforms is typified by password protection and smart card technology. Secure personal networks and services of the future require more; the establishment, maintenance, and ability to transfer trust are essential. It is not enough to create a secure end-to-end PN from one person or device to another; PNs and services must be shared. Establishing trust between friends may be relatively simple, but how can it be established between friends of friends or strangers? How can it be transferred? For example, exchange of information via personal networks at a conference. If we trust our friends and their devices and they trust their friends and devices, then access to our personal devices may quickly be open to all. If everyone is trusted, then no one is.

The determination of trust is not simple, and it is not the responsibility of the network and devices alone. It is also a system problem involving security of personal information and profiles that may be contained in our personal devices or in databases anywhere in the system. Although not a technical issue, it underscores the necessity of physical security as well. Shall our devices self-destruct in 5 seconds if they are accessed by an untrusted source?

The very complexity of the problem may impose additional requirements on the platforms or devices themselves. These are further complicated by the fact that many of these devices are resource limited, which means that low-complexity security mechanisms are needed. Nonetheless, it is the person in the loop that may pose the greatest risk. For example, simple copying of personal information in databases by workers inherently creates a security hazard. This is more of a trust killer than a technological barrier though. It suggests that a system-level security solution is required. The emphasis in this chapter is on flexible PAN platforms and devices for enabling practical, secure, end-to-end PNs.

6.9   SYSTEM OPTIMIZATION AND OPERATOR PERSPECTIVES

Personal networks may interact with a variety of systems in homes, vehicles, places of work, or sensor networks:

1. Home area networks (HANs)
2. Vehicular area networks (VANs)
3. Enterprise networks (ENs)
4. Wireless sensor networks (WSNs)

System optimization spans the user's P-PANs and personal devices and extends to the interaction of PNs with other systems and networks (e.g., HAN). Interaction with such systems may be needed in the future, not only to support communication but also to gather information concerning data for, for example, context and location-aware services.

The operator network enables wide-area connectivity to interconnect the various personal domains that may be geographically distributed. Such connections may be layer 2 or network-layer tunnels to ensure security in addition to device- and application-level security under the control of the user's own network domains. The operator can provide value-added services and additional capabilities in the network to offload the burden from the end user and charge for such offerings. Thus, the operator can serve the needs both in the horizontal or the vertical market segments and avoid becoming a mere bit-pipe provider.

6.10 TOWARD PERSONAL SERVICES OVER PERSONAL NETWORKS

Today, the market for personal services and personal networks remains relatively unexplored, in much the same way as it does more broadly for PAN and WSN. What we have seen may well be only the tip of the iceberg. The potential magnitude of the information traffic from the multitude of future personal and sensor devices is enormous, and so is the business potential, but specific applications and business cases remain to be determined. Last, but not least, if we are to support sensitive, even critical, personal information services, we need to build more than the networks, devices, and applications—we need to build the business and establish trust.

6.11 CONCLUSIONS

In this chapter, we described how a home network can evolve into a secured virtual home network, which may be distributed over a large geographic area. We defined such a network as a personal network (PN). A PN may consist of a personal area network (PAN), in turn, consisting of a short-range network interconnecting a myriad of personal devices that may move with the person, a home network, an office network, and trusted PNs of other users. We put forth a vision of a PN that would support the user's personal and private activities without being obtrusive and while safeguarding their privacy and security. We further showed by a number of example scenarios how PNs can be applied to broad application domains with a significant impact on the users' daily lives. We further developed a number of data exchange requirements, functional requirements, and overall system requirements for PNs and personal PANs. After a brief discussion on user requirements, network architectures, and access control techniques, we identified the security, privacy, and trusted relationships between the various entities of one's PN and those of other's PNs that it connects to as paramount for such networks before they can become viable and attractive for the users at-large. Nonetheless, although the market for personal services and PNs remains relatively unexplored, there is a huge interest in the global research community to tackle many of the challenges that such networks pose.

REFERENCES

1. IST MAGNET. Available at www.ist-magnet.org.
2. R. Tafazolli and J. Saarnio, Eds., eMobility Strategic Research Agenda, eMobility: Staying ahead! Available at http://www.emobility.eu.org/.
3. R. Prasad and K. Skouby, "Personal Network (PN) Applications." *Wireless Personal Communications*, vol. 33, nos. 3–4, 227–242, June 2005.
4. K. Vandrup, J. Farserotu and R. Prasad, MAGNET paving towards part for future wireless communications. EWCT'05, Paris, France, Oct. 2005.
5. I. Niemegeers and S. Heemstra de Groot, "From Personal Area Networks to Personal Networks a User Oriented Approach." *Journal on Wireless and Personal Communications*, vol. 22, no. 2, 175–186, 2002.
6. M. Jacobsson, et al., A Network Layer Architecture for Personal Networks. Available at http://www.ist-magnet.org/publications.html#WP2.
7. The Role and Impact of IPv6 in Implementing the PN, MAGNET D2.0, IST507102. Available at http://www.ist-magnet.org/publications.html#WP2.
8. Resource and Service Discovery: PN Solutions, MAGNET D2.2.1, IST507102. Available at http://www.ist-magnet.org/publications.html#WP2.
9. M. Ghader and R. Tafazolli, Performance of Service Discovery Protocols in Personal Networks. Available at http://www.ist-magnet.org/publications.html#WP2.
10. I. Niemegeers, S. Heemstra de Groot, "Research Issues in Ad-Hoc Distributed Personal Networking." *Wireless Personal Communications*, vol. 26, nos. 2–3, 149–167, August 2003.
11. Secure Wireless Personal Networks, The MAGNET PAN Air Interface Approach for Personal Networks—Access Techniques. Prepared by: J. Farserotu, F. Platbrood, R. Hoshyar, K. Schoo, C. Mutti, H. Choi, I. Siaud, J. Ayadi, J. Gerrits. Presented by: M. Presser, EWCT'04, Amsterdam, The Netherlands, 11–15 October 2004.
12. M. Presser, R. Tafazolli, I. Kovacs, D. Dalhaus, F. Platbrood et al., "MAGNET 4G Personal Area Network Air-Interfaces for Personal Networks." *Proc. 13th IST Mobile & Wireless Communications Summit*, Lyon, France, pp. 169–175, June 2004.
13. J. Gerrits, J. Farserotu and J. Long, "UWB Considerations for My Personal Global Adaptive Networks (MAGNET) Systems, *Proc. ESSCIRC* 2004, pp. 45–56, Sept. 2004.
14. J. Farserotu, et al., UWB Transmission and MIMO Antenna Systems for Nomadic Users and PANs. Special Issue on Designing Solutions for Unpredictable Future. *Wireless Personal Communication Journal*, vol. 22, no. 2, 297–317, August 2002.
15. K. Nyberg and D. Sisalem, Establishing Security in Personal Area Networks. EWCT'04 Amsterdam, The Netherlands, 11–15 October 2004.
16. N. Prasad, Adaptive Security for Low Data Rate Networks. EWCT'04, Amsterdam, The Netherlands, 11–15 October 2004.
17. J. Saarnio and N. Prasad, Foolproof security measures and challenges within. *Wireless Personal Communications* vol. 29, nos. 1–2, 101–108, 2004.

7

USABLE SECURITY IN SMART HOMES

Saad Shakhshir and Dimitris Kalofonos

Traditionally, computer security has been a high priority only for entities dealing with extremely sensitive data, such as the military, the government, and banks. Information technology (IT) experts in each individual branch or institution were able to configure and manage their own internal security mechanisms. As consumer devices become smaller and more powerful, however, average users are becoming increasingly reliant on these devices for storing and relaying sensitive information—such as bank account and contact information.

In tandem with the extensive proliferation of portable devices, there has been a considerable increase in connectivity. Today, there are more than 350 million hosts on the Internet [1]. Additionally, wireless networks are being set up in homes, offices, cafés, and malls thereby increasing interconnectivity even further. In fact, the number of home wireless networks in the United States reached 8.7 million in 2004. That number is expected to increase rapidly in the near future, and analysts estimate that there will be 28 million home wireless networks in 2008 [2]. At the same time, however, the lack of security is staggering. Estimates vary, but most analysts seem to agree that "between 60% and 70% of all existing wireless networks—corporate and personal—have no external security at all." In fact, the act of snooping around for unprotected wireless networks has become popular enough for it to merit a special name—wardriving [3].

One factor that substantiates the large projected increase in home wireless networks is the introduction of new network-capable household products. Microwaves, refrigerators, TVs, and almost any other consumer product will soon have some form of network connectivity thus transforming the average home into what

Technologies for Home Networking. Edited by Sudhir Dixit and Ramjee Prasad
Copyright © 2008 John Wiley & Sons, Inc.

is known as the "smart home." There are various initiatives currently working on standardizing the framework for interoperable device interaction within the smart home. For an example of such an initiative, see the Digital Living Network Alliance (DLNA) [4].

The emergence of the smart home coupled with the increased reliance on portable devices to handle sensitive information is ushering in a new status quo. This requires a paradigm shift in the design of secure systems. The average consumer cannot be expected to interact with security in the ways that IT experts traditionally have. These systems are not designed with nonexpert users in mind. Furthermore, consumer products within the smart home will communicate over several different wired and wireless media, such as Bluetooth, 802.11, Ethernet, and so forth. The varied nature of the underlying network medium only exacerbates the burden of security on the nonexpert user.

In this chapter, we discuss the introduction of a usable security framework in the smart home environment. We present a survey of related work on home network security and usability, and we present several of the most interesting basic use cases. We discuss the challenges that nonexpert users face when configuring security for their smart homes and setting up multiple devices to connect to the smart home network. We also discuss the difficulties users encounter when granting temporary access to visitors for specific services in the home. We then outline several approaches to creating an intuitive smart home security framework and discuss the advantages and disadvantages of each.

This chapter is organized as follows: Section 7.1 presents a survey of related work; Section 7.2 introduces some basic use cases for smart home security; and Section 7.3 describes the smart home security and threat model that we consider. Section 7.4 introduces the challenges of creating an intuitive smart home security framework; Section 7.5 discusses several different approaches to creating a usable security framework; finally, we summarize and conclude this chapter with Section 7.6.

7.1 SURVEY OF RELATED WORK

Smart home security includes research from three main disciplines: security, human–computer interaction (HCI), and smart spaces. Although there are many publications in each field individually, it is more meaningful and relevant for the purposes of this chapter to look at work that fits into some combination of the three.

7.1.1 User Interaction with Security

One approach in this domain has been to critically examine the usability of existing applications and technologies with respect to security. In Ref. 5, Whitten and Tygar perform a rigorous analysis on Pretty Good Privacy (PGP) and conclude that although it was intended to provide a global framework for securing and authenticating e-mails, it did not meet their usability requirements. A study was conducted on users in which they found that only 3 out of 12 test subjects successfully managed to use PGP to encrypt and decrypt e-mail. Similar conclusions are reached in Ref. 6 where Zurko et al. examined user behavior when presented with security question pop-ups during workflow while using the Lotus Notes application. They conclude that almost half of the users make the "wrong choice" and let the insecure content run. The Kazaa peer-to-peer (P2P) file-sharing application is looked at in Ref. 7 and found to fail in its handling of privacy-related functionality. Users inadvertently share information and files that they would like to keep private.

Other researchers have focused on more general aspects of security usability rather than examining specific applications. In Ref. 8, Dourish et al. conducted a series of qualitative interviews at both an academic institution and an industrial research lab. They found that "neutral to negative attitudes dominated the end-user experience of security technologies" and although not focusing on ubiquitous computing as such, they recognize the increased complexity and urgency of creating usable security frameworks for pervasive computing. Yet more papers examine the relationship between usability and security in an attempt to create general design guidelines [9, 10].

Another point that researchers make is the importance of presenting the user with appropriate visualizations in order to make informed decisions related to security [11]. DiGioia and Dourish take a novel approach in this regard leveraging social navigation to present more meaningful visualizations to the user [12]. Finally, there is some work that presents more concrete attempts at creating applications with a more usable interaction with security, but these are more recent. One such example is SPARCLE, a "set of privacy utilities that are intended to assist organizations with the creation, implementation, and internal auditing of privacy policies" [13].

Besides the above academic approaches, the issue of security usability has attracted a lot of interest in the industry, especially in the domain of wireless networking. There are two main initiatives worth noting: Buffalo Technology's AirStation OneTouch Secure System (AOSS) [14] and Linksys' SecureEasySetup (SES), which was adopted from Broadcom [15]. Both technologies promise quick and easy setup of secured wireless networks and have similar user interfaces despite using different underlying security mechanisms. Currently, there is an effort by the Wi-Fi Alliance to standardize this process using nonproprietary means [16].

7.1.2 Security in Smart Spaces

Again in this field there are several papers that give an overview of current research [17, 18]. Campbell et al. take this one step further in Ref. 19 by also presenting their own prototype implementation to address some of these issues. One approach to security in smart spaces is to use a set of environment variables to create an environment role that then dictates the access control to various resources [20]. Another framework has been proposed to manage key distribution at the link level [21]. Both these approaches use a centralized server.

At this point, it is important to describe security work done by the Universal Plug and Play (UPnP) Forum [22–25], which is directly applicable to home network security. The forum consists of a large number of companies across a wide range of technology industries and is strongly supported by DLNA, which has adopted their standards for interoperability within the smart home. The security aspect is one component of the larger UPnP Device Architecture that was standardized and adopted by the forum in 2003. The security framework proposed by the forum is specific to the smart home environment. It introduces a Security Console, which is a software component that handles the "client-side" of the security protocols. These Security Consoles are the identity-bearers in the home and not users. The Device Security component handles the "server-side" of the security interaction. The framework uses existing notions of certificates to grant access to devices. It, however, does not account for granting of connectivity access and the different connectivity mechanisms that a device may use to join a home network. Their proposals also do not focus on the framework's usability, which is clearly a key element of any smart home network infrastructure. A good overview of this work can be found in Ref. 26.

7.1.3   User Interaction with Security in Smart Spaces

This area combines research from all the above fields. In Ref. 27, the authors present a framework that uses trust levels in order to establish access control. Users are assigned levels of trust by the appliances. The description of how this is done is quite high-level, and it does not address security on the lower connectivity levels.

Others leverage location-limited channels (LLCs) as part of their frameworks. In Ref. 28, Balfanz et al. describe their "Network in a Box," which uses the infrared port of a laptop to autoconfigure the laptop's wireless network security settings. Another illustrative example is given in Ref. 29, where LLCs are used to exchange public key information between a laptop and a printer in an airport lounge. This then allows the laptop to securely print to that specific printer over Secure Sockets Layer (SSL) or Transport Layer Security (TLS). They follow up on this research in Ref. 30, where they describe five usability lessons learned from conducting an experiment with two different methods of setting up a public key infrastructure (PKI) based wireless network. The first method involved users performing manual configuration of their devices and the second involved using the infrared method described earlier. Their results showed that it took users an average of 140 minutes to configure their devices manually and only 1 minute 39 seconds using the LLC-based autoconfiguration method. All the participants were computer literate, typically with a Ph.D. in computer science.

Holmstrom takes a different approach in Ref. 31 using the metaphor of a business card to delegate permissions between individuals. The example given there is granting your colleague permission to read your e-mail while you are away. The framework presented uses Simple PKI certificates as these do not rely on a hierarchy of certification authorities thereby addressing the scalability problems of traditional X.509 identity certificates.

7.2   BASIC HOME SECURITY USE CASES

This section presents several basic use cases for user interaction with security in the smart home. The scenario consists of a household with a married couple—Bob and Alice.

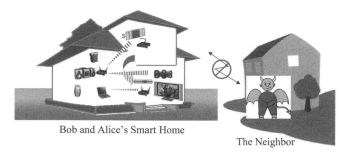

Bob and Alice's Smart Home

The Neighbor

FIGURE 7.1 Bootstrapping and securing the smart home network.

The first use case involves bootstrapping new devices into the smart home. When Bob and Alice purchase a new device, they would like it to have permanent connectivity to the smart home network over a secure channel. At the same time, they would like to protect their home from being accessed by nonauthorized users. Figure 7.1 illustrates this.

Once the devices are connected and can all communicate with each other securely, Bob would like to prevent Alice from accessing his devices until he explicitly grants her access. He could also opt to have his devices have some default level of access to everybody. By way of example, Bob purchases a new media server whose content he wishes to keep private. Both Bob and Alice jointly purchased the A/V renderer in the living room so they both have access to it by default. Thus, Bob can now stream content from his media server and display it on the A/V renderer, however Alice cannot. Figure 7.2 illustrates this.

Alice later decides that she wants to retrieve content from the server and asks Bob to give her access. Bob agrees, but he only wants to give her permission to download music and movies from the server. He does not want her to access any other file types

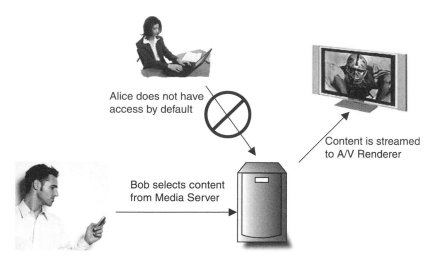

Alice does not have access by default

Content is streamed to A/V Renderer

Bob selects content from Media Server

FIGURE 7.2 Before Bob grants Alice access.

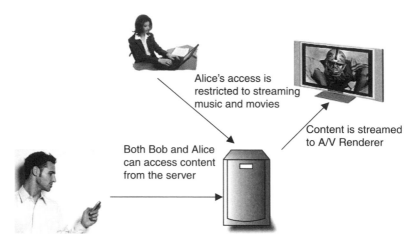

FIGURE 7.3   After Bob grants Alice access.

or to upload or delete anything. Figure 7.3 illustrates the situation after Bob grants Alice limited access to his media server.

A further use case involves granting temporary access to visitors for use of specific services in the smart home. Again, by way of example, the refrigerator in Bob and Alice's home breaks down. The repairman comes over and by default is not even able to connect to and browse their home network. However, Alice grants the repairman access for the day to selected functionality provided by the refrigerator so that he can perform the necessary repairs. At the end of the day, once the work is complete, the repairman automatically loses all access to Bob and Alice's smart home network. Figure 7.4 shows the repairman's access during the day.

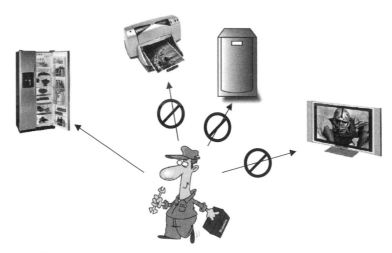

FIGURE 7.4   The repairman's access while repairing the refrigerator.

7.3 A SMART HOME SECURITY MODEL

In this section, we describe a typical smart home environment and the security model and threats that we assume. An example of a smart home is a house with both wired and wireless connectivity that contains numerous devices providing various services to users in the home. Such devices could be media servers, audio/visual (A/V) renderers, refrigerators, or any household appliance or consumer electronics device, as depicted in Figure 7.5. Currently, many of these devices are not yet network-capable, but the technology to make them so is available, and there are various initiatives strongly advocating for such adoption in the near future. For an example of such an initiative, see the Digital Living Network Alliance (DLNA) [4].

Smart homes are likely to have several occupants. We assume that each occupant would like to protect the privacy of her own devices while being able to grant selective access to other individuals as she sees fit. At the home network level, we assume that the occupants would like to prevent unauthorized non–home occupants from gaining any access to the home network; this includes even connectivity access, such as the ability to acquire an IP address and passively read network traffic. We also assume that home occupants would like to be able to grant visitors access to devices within their homes for temporary periods of time.

Thus the threats that the security framework has to protect against are

- Unauthorized connectivity access to the home network through any means—Bluetooth, 802.11, and so forth.
- Unauthorized access to services on a device in the home by anyone that has not been granted permission to access it by the owner of that device.

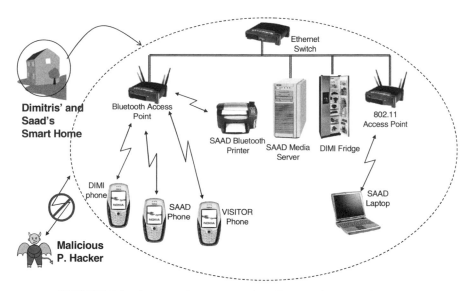

FIGURE 7.5 An example representation of a smart home environment.

To address these threats, access control must take place on two levels:

- Connectivity level: prevents unauthorized devices from connecting to the home network.
- Service/device level: prevents already connected users from accessing device functionality for which they are not authorized and allows users to reliably authenticate the devices they are interacting with.

In terms of protecting against unauthorized connectivity access, the smart home security framework needs to first focus on wireless connectivity. Gaining illicit wired access is seen as a much less significant threat to home network security than gaining access over a wireless connection, as the intruder must have physical access to the home in order to connect over a wire (i.e., they must break into a person's home before they can attach a malicious device with a wire). A device that is attached by wire is also more noticeable than a person snooping over a wireless connection and so could more easily be detected and removed. Furthermore, if an intruder does happen to acquire unauthorized wired connectivity to the home, he or she will still not be able to access the services offered by any of the home devices due to the second layer of protection at the service level. However, extending link-level security over the wired segment of the home network is also possible.

7.4 DESIGN CHALLENGES

There are several challenges that designers face when attempting to create a usable security framework for the smart home. The first of these challenges involves understanding and accommodating the multitude of different link-level security mechanisms of the connectivity options that are available, such as Bluetooth [32], Ultra Wide Band [33], Wireless USB [34], Wi-Fi [16], and WiMax [35]. Each of these has its own pool of security mechanisms such as Wired Equivalent Privacy (WEP) [36] or Wi-Fi Protected Access (WPA) for 802.11 [37], the pairing mechanism for Bluetooth [38]. Thus, any successful attempt at creating a smart home security framework will have to incorporate a variety of different wireless connectivity media and their respective link-level security mechanisms.

Second, once connections are secure at the link level, the security framework must handle authentication and access control at the device or service level. Furthermore, it is possible that some applications may require end-to-end encryption mechanisms, for example, based on the Secure Socket Layer (SSL), to protect the privacy of data exchanges between nodes in the smart home. Looking in more detail at the repairman scenario, in order for him to access the refrigerator, he must be given permissions at both the link level to connect to Alice and Bob's smart home network and at the device level to access the refrigerator. The challenge here is successfully separating the policy from the mechanism in such a way that access control can take place on any type of device. The underlying mechanisms may well vary on each device; however, the policy of access control should remain consistent.

Third and perhaps most importantly is the issue of usability. As the smart home incorporates an increasing number of devices, the complexity of the network will only increase making it less manageable for the average user. The issue of usability is one that needs to be addressed as part of all aspects of the smart home. The fact that currently around two thirds of wireless networks are not secure demonstrates how important this factor is when it comes to creating a security framework for the smart home. Even more experienced users face difficulties when configuring and managing their networked devices as in most cases they must interact directly with low-level security concepts, such as WEP keys and MAC address filters.

Finally is the issue of architecture. There are arguments for the creation of a central security device that manages the security of various other devices. This could become a requirement of the smart home network, and any attempt to connect to a device would first need to be approved by the central security device. However, there are strong industry initiatives pushing for the standardization of a decentralized approach where each device handles its own security policy [23].

7.5   USABILITY

There is a clear need to focus on usability when designing a security framework for smart homes. A lot of work has been done on developing the underlying security mechanisms and ensuring their cryptographic strength; however, until recently there has been little emphasis on creating a consistent and intuitive interaction between nonexpert users and security frameworks. Currently, most users are faced with the daunting task of dealing directly with low-level security parameters, most of which are cryptic even to the intermediate or advanced user.

For example, manufacturers of most wireless 802.11 access points currently provide users with an interface to configure the security settings of their devices over an HTTP connection. When a user first starts up their new access point, they are normally required to connect to it through a Web browser after which they are presented with a page such as in Figure 7.6. Terms such as WPA, WEP keys, ASCII, Hex, and 128-bit easily overwhelm even the relatively knowledgeable user.

In addition to link-level security settings, users must configure their devices to grant access to specific users. This usually involves manually editing some type of Access Control List (ACL) on the device. These entries are permanent and difficult to manage as users must continually revisit the ACLs in order to remove outdated entries and update existing ones. This process is cumbersome and not intuitive.

Taking the example of the refrigerator repairman, current security frameworks would require Bob or Alice either to perform some type of manual configuration on their access points, such as modifying a MAC address list, or to enter a key (for WEP) or PIN (for Bluetooth) into the repairman's device. This would then grant him link-level connectivity access to the home. They would then additionally have to modify the ACL of the refrigerator in order to grant the repairman access to whatever functionality is necessary for the repairs. Once the repairman has completed his repairs, they would have to go back and remove his entry from the ACL

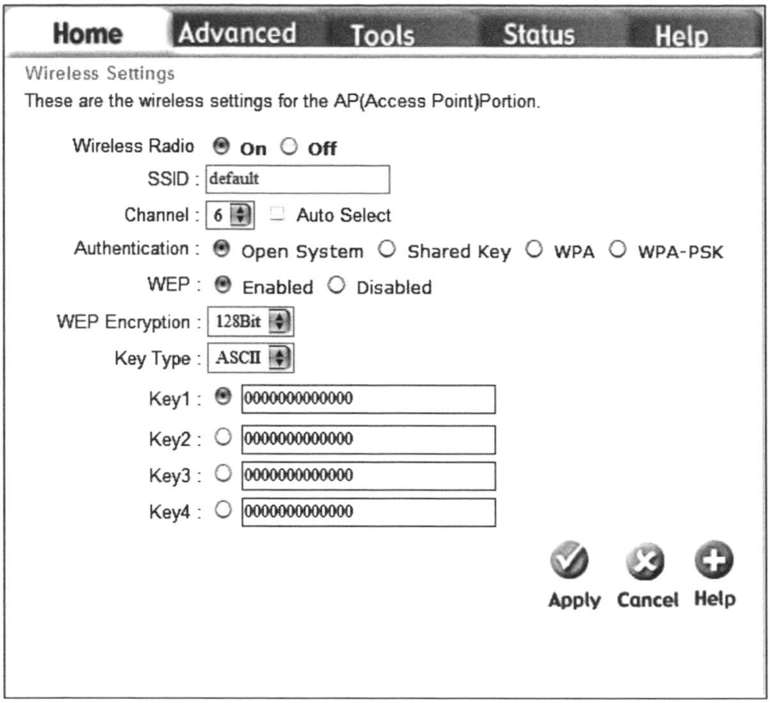

Home	**Advanced**	**Tools**	**Status**	**Help**

Wireless Settings

These are the wireless settings for the AP(Access Point)Portion.

Wireless Radio ◉ **On** ○ **Off**

SSID : default

Channel : 6 ⬍ ☐ Auto Select

Authentication : ◉ Open System ○ Shared Key ○ WPA ○ WPA-PSK

WEP : ◉ Enabled ○ Disabled

WEP Encryption : 128Bit ⬍

Key Type : ASCII ⬍

Key1 : ◉ 0000000000000

Key2 : ○ 0000000000000

Key3 : ○ 0000000000000

Key4 : ○ 0000000000000

Apply   Cancel   Help

FIGURE 7.6   A typical setup screen provided by an 802.11 wireless access point.

of the refrigerator. They would also need to change the link-level keys to make sure that the repairman cannot reconnect to their home network.

Realizing that the complexity of these steps places an unrealistic burden on the average user, there have been several attempts at creating a more usable interaction with security for home networks. Microsoft released a new wireless setup wizard as part of a recent upgrade to its Windows XP operating system. The wizard attempts to walk the user through the creation of a secure wireless network using several relatively simple steps. A screenshot of this wizard is shown in Figure 7.7.

Two other initiatives worth noting are Buffalo Technology's AirStation OneTouch Secure System (AOSS) and Linksys' SecureEasySetup (SES), which was adopted from Broadcom. Both technologies promise quick and easy setup of secured wireless networks and have similar user interfaces despite using different underlying security mechanisms. AOSS supports security levels from the relatively weak 64-bit WEP to the strongest available WPA2-PSK, and SES supports devices capable of WPA-PSK/TKIP security. An image of the Linksys WRT54G wireless router that supports SES is shown in Figure 7.8. After pressing the Cisco logo on the router, the software then automatically detects and creates a secure connection with the router.

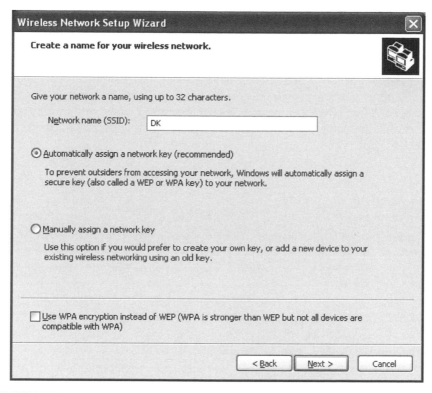

FIGURE 7.7   The wireless network setup wizard as part of Windows XP Service Pack 2.

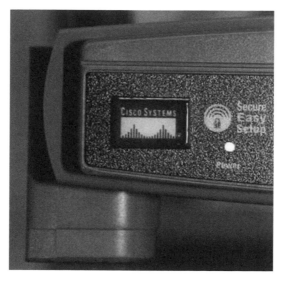

FIGURE 7.8   Linksys WRT54G wireless router supporting SecureEasySetup.

7.6 CONCLUSIONS

In this chapter, we presented an overview of issues pertaining to the development of usable security frameworks for smart homes. There is a clear need for the development of such frameworks, and steps are already being made toward this goal, as indicated by the various academic and industry initiatives. However, much work remains to be done. A framework that incorporates the majority of underlying security mechanisms while presenting the user with a consistent and usable interaction has yet to be achieved.

REFERENCES

1. Internet Systems Consortium Internet Domain Survey. Available at http://www.isc.org/index.pl?/ops/ds/.

2. R. Lieb, "Wi-Fi Moves In." Available at http://www.clickz.com/stats/sectors/wireless/article.php/3416331.

3. C. Hurley, M. Puchol (Editor), R. Rogers, F. K. Thornton, *WarDriving: Drive, Detect, Defend, A Guide to Wireless Security.* 1st ed., Syngress, 2004.

4. Digital Living Network Alliance (DLNA), "Home Networked Device Interoperability Guidelines v1.0," June 2004. Available at http://www.dlna.org.

5. A. Whitten and J. Tygar, "Why Johnny can't encrypt: A usability evaluation of PGP 5.0." In *Proceedings of the Eighth USENIX Security Symposium*, Washington, DC, 1999.

6. M. Zurko, C. Kaufman, K. Spanbauer, and C. Bassett "Did you ever have to make up your mind? What notes users do when faced with a security decision." In *Proceedings of the 18th Annual Computer Security Applications Conference (ACSAC)*, Las Vegas, NV, 2002.

7. N. S. Good and A. Krekelberg, "Usability and privacy: A study of kazaa p2p file-sharing." In *Proceedings of the Conference on Human factors in Computing Systems (CHI'03)*, p. 137–144, ACM Press, Ft. Lauderdale, 2003.

8. P. Dourish, R. E. Grinter, J. Delgato de la Flor, and M. Joseph, "Security in the wild: User strategies for managing security as an everyday, practical problem." *Personal and Ubiquitous Computing*, 8(6), 2004, 391–401.

9. K.-P. Yee, "User interaction design for secure systems." In *Proceedings of the 4th International Conference on Information and Communications Security (ICICS)*, Singapore, December 2002.

10. M. E. Zurko and R. T. Simon, "User-centered security." In *Proceedings of the New Security Paradigms Workshop*, Lake Arrowhead, CA, September 1996.

11. R. De Paula, X. Ding, P. Dourish, K. Nies, B. Pillet, D. Redmiles, J. Ren, J. Rode, and R. Silva Filho. "Two experiences designing for effective security." In *Symposium On Usable Privacy and Security (SOUPS 2005)*, Pittsburgh, PA, 2005.

12. P. DiGioia and P. Dourish, "Social navigation as a model for usable security." In *Proceedings of the 2005 Symposium on Usable Privacy and Security*, pp. 101–108. Pittsburg, PA, ACM, 2005.

13. C. Brodie, C. M. Karat, J. Karat, and J. Feng, "Usable security and privacy: A case study of developing privacy management tools." In *Proceedings of the 2005 Symposium on Usable Privacy and Security*, pp. 35–43. Pittsburg, PA, ACM, 2005.

14. Buffalo Technology. AirStation One-Touch Secure System (AOSS™), October 2004. Available at http://www.buffalotech.com/documents/pdf/AOSS_WP_Final.pdf.

15. Broadcom Corporation, Securing Home Wi-Fi Networks: A Simple Solution Can Save Your Identity, May 2005. Available at http://www.54g.org/pdf/Wireless-WP200-RDS.pdf.

16. Wi-Fi Alliance (WFA). Available at http://www.wi-fi.org.

17. J. Wang, Y. Yang, and W. Yurcik, "Secure smart environments: Security requirements, challenges and experiences in pervasive computing." In *Experience Workshop on Pervasive Computing*, Urbana, IL, July 2005.

18. P. A. Nixon, W. Wagealla, C. English, and S. Terzis, "Security, privacy and trust issues in smart environments." In *Smart Environments: Technologies, Protocols and Applications*. Wiley-Interscience, Hoboken, NJ, 2005.

19. R. Campbell, J. Al-Muhtadi, P. Naldurg, G. Sampermane, and M. D. Mickunas, "Towards security and privacy for pervasive computing." In *Proceedings of the International Symposium on Software Security*, Tokyo, Japan, 2002.

20. M. J. Covington, W. Long, S. Srinivasan, A. K. Dey, M. Ahamad, and G. D. Abowd, "Securing context-aware applications using environment roles." In *Proceedings of the ACM Symposium on Access Control Models and Technologies*, Chantilly, VA, May 2001.

21. H. Nakakita, K. Yamaguchi, M. Hashimoto, T. Saito, and M. Sakurai, "A study on secure wireless networks consisting of home appliances." *IEEE Transactions on Consumer Electronics*, pp. 375–381, May 2003.

22. UPnP Forum. Available at www.upnp.org.

23. UPnP Forum, "UPnP security ceremonies v1.0." October 2003. Available at http://www.upnp.org/download/standardizeddcps/UPnPSecurityCeremonies_1_0secure.pdf.

24. UPnP Forum, "UPnP device security v1.0." November 2003. Available at http://www.upnp.org/standardizeddcps/documents/DeviceSecurity_1.0cc_001.pdf.

25. UPnP Forum, "UPnP security console v1.0." November 2003. Available at http://www.upnp.org/standardizeddcps/documents/SecurityConsole_1.0cc.pdf.

26. C. Ellison, "Home network security." *Intel Technology Journal*, 6(4), 2002, 37–48.

27. J. Seigneur, C. Jensen, S. Farrell, E. Gray, and Y. Chen, "Towards security autoconfiguration for smart appliances." In *Proceedings of the Smart Objects Conference*, Grenoble, France, 2003.

28. D. Balfanz, G. Durfee, R. Grinter, D. Smetters, and P. Stewart, "Network-in-a-Box: How to set up a secure wireless network in under a minute." In *Proc. of 13th USENIX Security Symposium*, San Diego, CA, 2004, pp. 207–222.

29. D. Balfanz, D. Smetters, P. Stewart, H. Chi Wong, "Talking to strangers: Authentication in ad-hoc wireless networks." In *Proceedings of the Network and Distributed Systems Security Symposium (NDSS'02)*, San Diego, CA, 2002.

30. D. Balfanz, G. Durfee, D. K. Smetters, "In search of usable security: Five lessons from the field." *IEEE Security & Privacy*, pp. 19–24, September/October 2004.

31. U. Holmstrom, "User-centered design of security software." In *Proceedings of 17th International Symposium on Human Factors in Telecommunications*, Copenhagen, Denmark, May 1999.

32. Bluetooth Special Interest Group, "Bluetooth Core." Specification of the Bluetooth System version 1.2, November 2003.

33. WiMedia Alliance. Available at http://www.wimedia.org.

34. Wireless Universal Serial Bus Specification 1.0, May 2005. Available at http://www.usb. org/wusb/docs/WirelessUSBSpecification_r10.pdf.

35. WiMax Forum. Available at http://www.wimaxforum.org.

36. ANSI/IEEE 802.11,"802.11std. Wireless LAN Medium Access Control and Physical Layer specifications." August 1999.

37. Wi-Fi Alliance; "Wi-Fi Protected Access (WPA)." October 2002.

38. Bluetooth Special Interest Group, "Bluetooth Security Architecture." White paper, version 1.0, 15 July 1999.

8

MULTIMEDIA CONTENT PROTECTION TECHNIQUES IN CONSUMER NETWORKS

HEATHER YU

In the past several years, content protection (CP) has been a media buzzword. Digital multimedia content protection at home is an area of interest for content providers, consumer electronics (CE) companies, and information technology (IT) companies for various reasons. Primarily, protection of multimedia art works from unauthorized distribution or modification is a key concern for content providers. The Motion Picture Association of America (MPAA) estimates that the U.S. motion picture industry loses in excess of $3 billion annually in potential worldwide revenue due to piracy. The Recording Industry Association of America (RIAA) claims their industry loses about $4.2 billion per year to piracy.

In the past 20 years, as technologies in data compression, manipulation, management, and searching advances, and as the availability of low-cost and high-quality multimedia entertainment devices and the digital consumer device interconnectivity escalates, consumer multimedia content accessing, authoring, and handling are getting easier and easier. Today, with a <$100 portable MP3 player, a $200 iPod, a <$50 CD rewritable drive, or a <$100 DVD recordable drive, the average consumer can copy any digital audio, CD album, or DVD video and any song or movie downloaded from the Internet onto a portable MP3 player, an iPod, or a $0.20 CD-R or DVD-R disk. Alternatively, it can be placed on a Digital Video Recorder (DVR) or a home computer set to play at any time or on a Web site ready for the 1 billion Internet population to access and download freely. What's more? Consumers can easily stream a video from a PC or a digital camcorder to a TV in

Technologies for Home Networking. Edited by Sudhir Dixit and Ramjee Prasad
Copyright © 2008 John Wiley & Sons, Inc.

a digital home. In the not too distant future, consumers will be able to redirect any multimedia stream from one device to another interconnected via consumer networks. This new exciting set of capabilities are making consumers increasingly interested in multimedia communications with transparent home networking for entertainment and information. Consequently, creating attractive products and services for home networked digital entertainment and information appliances is becoming an appealing field for both CE and IT industries. However, without proper content protection systems in place, it could potentially proliferate the motion picture and the recording industries' revenue loss. To balance the digital multimedia business opportunity and "copyright protection" with "fair use," many content providers are willing to define the home network as an acceptable boundary for "fair use" of multimedia content. Consequently, content providers, CE and IT industries joined the effort in defining multimedia content protection technologies. Now, many industry forums and standardization bodies have investigated or are investigating content protection approaches to satisfy various content providers' security requirements.

Other security concerns for home networked multimedia content include privacy and home security risks incurred by digital connectivity (i.e., the accessibility of home appliances via the Internet or other wide-area networks). In this chapter, however, we will focus on technologies for media security (i.e., media content protection only). Specifically, Section 8.1 gives a brief overview of current multimedia protection technologies in general. Several consumer networking specific content protection techniques are presented in Section 8.2, followed by issues and discussions in Section 8.3

8.1 TECHNIQUES FOR MULTIMEDIA CONTENT PROTECTION

8.1.1 Basic Security Requirements for Content Protection

8.1.1.1 Application Requirements Commercial and home-made multimedia content access applications (see Table 8.1) are most popular in today's home entertainment. Consumers may receive commercial content via various means, such as prepackaged media, broadcasting, and downloading or streaming through the Internet. Once receiving the content, the consumer may want to play back or store the content in the current device or transfer it to another consumer device on the home network.

Consumers may also create their own content. For privacy or other reasons, they may not want anybody outside their home to access the content. However, for convenience of usage, they may prefer to have access capability across their own home network.

From the application standpoint, specifically, consumers desire low-cost and low-complexity but high-quality devices and services. They care about privacy and security. But fewer would be willing to pay big bucks for a byzantine device or service that offers heightened security. Hence, ideally, a content protection system should balance information protection, usability, and cost to provide a beneficial environment for all parties involved.

TABLE 8.1 Sample Home Networked Multimedia Applications

Applications	Major Content Type	Key Security Concerns	Affected Party
Commercial content for entertainment	Video, audio, image	Unauthorized playback, unauthorized duplication	Content creator, content provider
Home-made content for entertainment	Video, audio, image	Unauthorized access, unauthorized duplication, unauthorized usage (misuse)	Owner of the content (consumer)
Private data for information	Text, image	Unauthorized exploitation, sunauthorized alteration	Owner of the data (consumer)

8.1.1.2 Technology Requirements Content protection technologies deal with methods and tools to prevent unauthorized access and distribution of multimedia content, for example, unauthorized playback or recording of a protected content, unauthorized transmission of content from one device to another, or illegitimate distribution of the content from one user to another. From a technology point of view, protecting media content requires the satisfaction of three goals to prevent dire consequences caused by malicious hackers [1], namely:

Integrity: to determine the authenticity of the content (i.e., if the content has been tampered with)

Confidentiality: to ensure privacy at transmission

Authorization: to control access and prevent theft of content (i.e., to protect it from illegal access and distribution)

8.1.2 Traditional Techniques

8.1.2.1 Encryption and Authentication Traditionally, data authenticity and confidentiality are addressed with cryptographic tools [2]. *Encryption* via asymmetric or symmetric key algorithms and *authentication* with one-way hash functions are commonly used to establish the integrity and privacy of data. Traditional cryptographic functions applied to multimedia are able to protect information from eavesdropping and detect information tampering. The two components required for data encryption are algorithm and key. The algorithm is usually known, but the key is kept secret. Figure 8.1 illustrates the idea and procedure of multimedia encryption.

8.1.2.2 Key Management Key management usually refers to the handling of cryptographic keys during the entire life cycle of the keys, including their generation, establishment, storage, transportation, usage, and destruction.

Key management is one of the most critical aspects of a cryptographic system such as an access rights management system for multimedia content distribution. This is

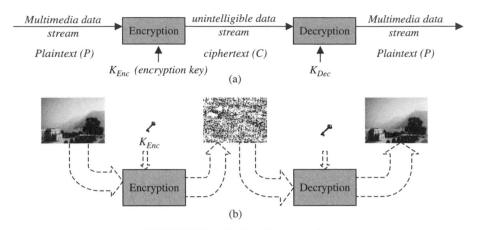

FIGURE 8.1 Multimedia encryption.

because the security measure in cryptography is based on the secrecy of the keys. Without the key, an attacker cannot successfully view, modify, or fabricate media content or information associated with them.

8.1.2.3 Challenges for Multimedia Applications

Due to the large size of multimedia data and different access requirement, conventional cryptographic tools are somewhat insufficient for today's applications. First of all, video and audio are megabytes to gigabytes in size. The additional computational power requirement for encryption makes it impractical for many applications, such as media streaming. Second, traditional encryption algorithms are sensitive to changes, making them unfeasible to protect format changeable and transcodable media content. Format change, commonly used in a wide range of applications, does not alter the meaning or originality of the multimedia content. For instance, a high definition (HD) video clip may be transferred via the home network to a device with different capabilities. One can imagine transferring the HD video onto a portable playback device will most probably trigger some compression or transcoding operation and cause media format change. In this kind of home networked applications, format change needs to be accepted under integrity and confidentiality checks. Third, traditional encryption algorithms do not address the issue of leakage; that is, once encryption is no longer active, the content is vulnerable to many kinds of unauthorized uses. De-Content Scrambling System (DeCSS) is a good example. DeCSS essentially decrypts DVD movies and facilitates their free storage and circulation without requiring royalty payments. From usability and quality of service points of view, limited random accessibility and scalability using conventional cryptographic tools, lack of scalable encryption keys and flexible key revocation schemes, and incompatible interoperability and renewability of different crypto systems also suggest the need of new technologies for multimedia content protection.

8.1.3 Advanced Cryptography Algorithms for Multimedia Content Protection

Studies on customizing conventional cryptography algorithms for multimedia applications have been conducted. For example, to save computational power (i.e., increase the encryption/decryption speed for media streaming applications), selective video encryption was proposed. For instance, Alice may want to stream a new video taken at a park to Mike's cellular phone in real time. Selective encryption is a straightforward way to reduce the overall computation time at both the source (digital camcorder) and the sink (cellular phone) devices and still preserve a certain level of security. Alice can have Mike instantly enjoying the video without worrying about it being maliciously modified in transmission. In Ref. 3, Maples and Spanos proposed to encrypt only the I-frame of MPEG video stream as there is a dependency of B- and P-frames to I-frames. Many other selective encryption algorithms have been proposed in the literature. (Interested readers can see Ref. 4 for a comprehensive survey on selective encryption techniques.) The fundamental idea is to partition the data into segments and select only the content-sensitive segments to encrypt. Without losing generality, one can partition a media stream in the other domains and selectively encrypt some segments but not all for speedy encoding and decoding. The design of such a system should be based on the security requirement to balance the trade-off between security level and complexity.

8.1.4 Digital Watermarking

To solve the leakage problem, a few years ago a new technology known as digital watermarking was proposed. DVD video and DVD audio are among the first to adopt digital watermarking for media security including *copyright protection* and *access control*. Numerous conferences and workshops, such as the Information Hiding Workshop sponsored by the International Federation for Information Processing (IFIP), the Multimedia Security Workshop at the Association for Computing Machinery (ACM) Multimedia, the Security and Watermarking of Multimedia Contents Conference at The International Society for Optical Engineering (SPIE) Photonics West, and the sessions at the Institute of Electrical and Electronics Engineers (IEEE), Information Technology: Coding and Computing (ITCC), International Conference on Multimedia and Expo (ICME), International Conference on Image Processing (ICIP), and International Conference on Acoustics, Speech, and Signal Processing (ICASSP), evidence the effort.

Digital watermarking [5] is the process to embed a discreet data stream imperceptibly into a digital medium data stream using steganographic (i.e., data hiding [5]) principles. Let's denote $\overset{p}{=}$ perceptual equality. If $A \overset{p}{=} B$, A and B are perceptually no different. A digital watermarking system is a six tuple $(\mathcal{I}, \mathcal{I}', W, \mathcal{K}, E, D)$, where the following conditions are satisfied: (1) $\mathcal{I}$ is a finite set of host data streams; (2) $\mathcal{I}'$ is a finite set of possible watermarked data streams; (3) $\mathcal{K}$ is a finite set of possible keys; and (4) W is a finite set of possible watermarks with watermarking encoding and decoding rules $Enc_K \in E$ and $Dec_K \in D$ such that

$$Dec_K(I', I, w) = Dec_K(Enc_K(I, w), I, w) = w$$
$$\text{or } Dec_K(I', I) = Dec_K(Enc_K(I, w), I) = w$$
$$\text{or } Dec_K(I', w) = Dec_K(Enc_K(I, w), w) = w$$
$$\text{or } Dec_K(I') = Dec_K(Enc_K(I, w)) = w$$

$$\text{and } I' \overset{P}{=} I$$

where $I \in \mathcal{I}$ is an origxinal (host) media data stream; $w \in W$ represents the watermark, which is the data stream to be hidden; $I' \in \mathcal{I}'$ refers to the overall composite (watermarked) data stream; and $K \in \mathcal{K}$ is the key. In many applications, such as electronic media distribution, it should be possible to extract or detect the embedded watermark without any reference to the host signal. That is, $w = Dec_K(I')$. For content protection applications, fragile watermark [6], one that is readily modified with any alteration on the watermarked media, can be used for content *authentication* and *tamper proofing*. Robust watermark [7], intended to endure common signal processing and some level of intentional attacks, on the other hand was proposed to address the leakage problem. For instance, an identification code *ID* or access control code *C* may be mapped to a watermark w_r, $w_r = f(ID)$ or $w_r = f(C)$, and robustly embedded into a host medium to actively or passively manage copyright or access rights. More recently, semifragile watermark, one that can survive certain kinds of attacks, was invented for media authentication that is expected to actively locate and recover the tampered data [8]. Figure 8.2 illustrates the idea and a general procedure of digital watermarking. Detailed description of digital watermarking technology can be found in Ref. 5.

Unlike encryption, a robust or semifragile digital watermark w_r is designed to have substantially less sensitivity over compression modification and transcoding. When a

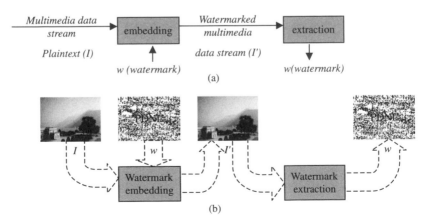

FIGURE 8.2   Multimedia watermarking.

content invariant feature $Q = F(I)$ is used, authentication value h can be generated based on Q, $h = H(Q)$, instead of the original data stream I. Providing the watermark's robustness over compression and transcoding, $I' = Enc_K(I, w_r)$, where w_r is a function of h: $w_r = f(h)$. $h^* = h$ maybe extracted from $I'' = I' + n$ after compression and transcoding transformation that imposes noise n. Because Q is designed to be content invariant under common linear or nonlinear transformation, $Q' = F(I'') = Q$, hence $h' = H(Q') = H(Q) = h$. An *adaptive content verification*, if $h' = h^*$, can be subsequently done.

For content *copyright protection*, one anticipated advantage of using digital watermark is to improve the robustness of copyright data against removal attacks. Robust watermark can be used to have the copyright data strongly associated with the host media, making it hard to remove. Coupled with cryptography tools, the hidden copyright data secrecy is expected to improve.

8.2 TECHNIQUES FOR CONTENT PROTECTION IN CONSUMER NETWORKING ENVIRONMENT

8.2.1 Existing Consumer Entertainment Content Protection Technologies: A Quick Overview

The major types of contents used in the consumer networking environment include standard-definition and high-definition audio, image, and video over broadcast channel, streaming through the home network, on prerecorded and recordable media and on high-definition prerecorded and recordable media. Currently available content protection technologies cover both commercial and personal content protection with an emphasis on commercial high-value multimedia content but mostly without fair use consideration under home networked environment. For example, the Content Scramble System (CSS) for DVD content, the Content Protection for Recordable Media and PreRecorded Media (CPRM/CPPM) for prerecorded/recordable media content protection, and Broadcast Flag for broadcasting content protection are some of the most frequently talked about technologies. Content encryption and device authentication techniques are employed by almost all existing content protection systems for the protection of consumer entertainment contents. In the following, let's briefly look at some of those proposed or adopted technologies. Table 8.2 summarizes a list of these technologies.

The CSS [9] is one of the earliest techniques developed for consumer entertainment content protection. It is an encryption system, with a proprietary, weak 40-bit encryption stream cipher, used on many DVDs. It was found in 1999 that CSS is susceptible to a brute force attack and the birth of DeCSS makes all protected DVD content easily accessible.

CPRM and CPPM [10] were invented to define a renewable cryptographic method that utilizes Cryptomeria cipher (C2) for protecting entertainment content when recorded on physical media. The types of physical media supported include the popular recordable DVD media and Flash memory.

TABLE 8.2   A Brief List of Popular Technologies

Technology	Target for Protection	Developed/ Developing By
AACS	From audio to video to high-definition video on various optical media formats including next-generation prerecorded and recorded optical discs	AACS LA
BDCPS	Multimedia content on BluRay Disc	3C
Broadcast flag	Broadcast content	CPTWG
CPPM	Entertainment content on prerecorded physical media such as DVD	4C
CPRM	Entertainment content on recordable physical media such as DVD-R	4C
CPSA [15]	A system framework for integration of various content protection techniques	4C
CSS	Content on DVD	DVDCCA
DTCP	Audio and video entertainment content transmitting through digital buses	5C
DVD audio watermark	Audio recorded on DVD	4C/Verance
DVD video watermark	Video recorded on DVD	CPTWG
HDCP	Digital entertainment content across the DVI/HDMI interface	Intel and Digital-CP LLC
Marlin [16]	Multimedia content shared on different devices and services across home network or consumer domain	Marlin Developer Community

To protect broadcast content, Broadcast Flag (BL) [11] was proposed by the Copy Protection Technical Working Group (CPTWG). BL is a set of status bits (or "flags") set in the data stream of a digital broadcasting program that indicates whether or not it can be recorded and if there are any restrictions on the recorded content, such as the rights to save it to a hard disk or other nonvolatile storage, the rights to make secondary copies of recorded content for sharing or archiving, the allowable quality when recording, and the rights to skip over commercials. Because many have asserted that broadcast flags interfere with consumers' fair use rights, BL has not been approved to date.

Most existing technologies for consumer content protection do not consider content access and distribution over an interconnected consumer environment. Digital Transmission Content Protection (DTCP) [12] is one of the earliest technologies for content protection over such environment. It defines a cryptographic protocol for protecting audio/video entertainment content from illegal copying, intercepting, and tampering as it traverses high-performance digital buses, such as the IEEE 1394 standard. It relies on strong cryptographic technologies to provide flexible and robust copy protection across digital buses. Only legitimate entertainment content delivered to a source device via another approved copy protection system (such as the DVD Content Scrambling System) will be protected by this copy protection system.

High-bandwidth Digital Content Protection (HDCP) [13] is a specification developed by Intel Corporation to protect digital entertainment content across the Digital Visual Interface (DVI)/High-Definition Multimedia Interface (HDMI) interface. The HDCP specification provides a robust, cost effective, and transparent method for transmitting and receiving digital entertainment content to DVI/HDMI-compliant digital displays. HDCP includes the set of authentication, encryption with stream cipher, and key revocation tools.

A more recent proposal, the Advanced Access Content System (AACS) [14], is developed for managing content stored on the next generation of prerecorded and recorded optical media for consumer use with PCs and CE devices. It is developed based on the broadcast encryption and uses Advanced Encryption Standard (AES) with title-specific decryption keys. The AACS protocols aim at allowing the content to be transferred to portable and networked devices and will interoperate with existing DRM (Digital Rights Management) schemes. It provisions consumer freedom to pipe content throughout the home, and, in some cases, to portable devices. For instance, a movie on a DVD could be ripped to a hard drive without the need to necessarily play the disks themselves. AACS also intends to interact with other DRM schemes, such as content protection within a network.

8.2.2 The Consumer Network "Boundary Problem"

In recent years, many raised the question of balancing content protection with user convenience. Consumers are skeptical about consumer devices that implement content protection technologies and are frustrated with the many content protection rules imposed on the use of some purchased content. For example, some companies, such as Microsoft and Adobe, use fixed number of devices as the constraint for software content protection to maintain equilibrium between the two opposing components of content protection—fair use and copyright. This kind of solution does not attempt to tackle the problem of content protection on interconnected devices. Further, it causes inconvenience when consumers wish to use the content on a different device.

In an interconnected consumer environment, a customer may want to play a music album that is loaded in computer X in the family room at computer Y upstairs in the bedroom. If the content protection rules restrict such type of "sharing" or "accessing," convenience and freedom of use is greatly limited. To deal with this problem, content company proposes to "localize" the transmission of content to home or personal network, that is to restrict the access to digital multimedia content within the home and personal network. For this reason, localization limitation, to limit the content distribution within a home or personal network and restrict protected content transmission and access across home or personal networks, is referred to as "boundary problem."

In response to the content provider's request, DTLA (Digital Transmission Licensing Administrator) defined a RTT (Round-Trip Time; the round-trip latency between a source and a sink device) [17] mechanism that permits the exchange of content across foreseeable home network. The idea is to use RTT as a measurement to control the use of the content. When the RTT is smaller than a predefined upper bound, the source and the sink devices are considered to be on the same home or personal network; content may be shared for user convenience. When the RTT is larger than the predefined upper bound, the source and the sink devices will not be recognized as sitting in the same home or personal network. Therefore, content protection rules will be imposed. Noticeably, although the above proposed mechanism is simple to implement with interoperability to many existing systems and can provide some user convenience, it cannot offer adequate balance between content protection and user convenience, as RTT could hardly provide accurate measurement for the targeted application. Thus, new technologies that can overcome the limitations of RTT are needed.

xCP (eXtensible Content Protection) by IBM takes another approach. It builds on the existing CPRM (Content Protection for Recordable Media) and CPPM (Content Protection for Pre-Recorded Media) [10] digital rights management technologies and requires little or no Internet connectivity and works by allowing all of the devices within a "home" network to establish common media keys. The unique IDs distinguish each device in the network from other devices, preventing use (viewing or listening) of the content by devices outside the network. Seemingly, it allows "fair use" mobility within a household network and, at the same time, prevents unauthorized digital content distribution from one household to another. To achieve this, a key has to be assigned and embedded to each and every device in a household. To access protected content, a user has to get compliant devices in his or her "consumer network." The device manufacturer needs to obtain the keys to be embedded from the centralized key issuer and manager. A content provider will also need to get a certified key from the key issuer. Because one key can unlock the content on all devices within the "home network," limited scalability is offered. A content provider may not define limited usage of certain content otherwise.

To provide better localization limitation, the rights object should be made home network or customer dependent instead of device dependent; that is, the rights object should be associated with the targeted home network instead of a particular device on the home network to ensure content can be shared within the home network for customer convenience. It should also be made scalable to the consumer network topology as well as to usage rules defined by the content providers.

8.2.3   Case Study: Protecting Streaming Media in Heterogeneous Network Environment

Noticeably, home network environment is highly heterogeneous (Fig. 8.3.) The varying network bandwidth coupled with the network dynamics, the diverse device capabilities, such as processing power and storage space, adds another level of difficulty for multimedia content protection at networked home environment.

8.2.3.1   An Application Scenario   Let's consider the following consumer application scenario. A consumer purchased a streaming video to stream to several

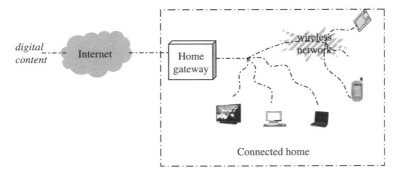

FIGURE 8.3   Heterogeneous devices in a connected home.

devices, such as DTV, desktop PC, PDA (personal digital assistant), and cellular phone, on the home network.

8.2.3.2 Scalable Plaintext Media Streaming   To meet the scalability requirement for audio and video streaming, simulcast as well as multiple description coding (MDC), scalable and fine-grained scalable (FGS) compression algorithms were proposed to provide scalable access of multimedia with interoperability between different services and flexible support to receivers with different device capabilities. With simulcast, multiple bitstreams of multiple bit-rates for the same content are generated. Unlike simulcast, scalable coding and MDC based schemes do not require the server to compute and save N copies of the source medium data stream. With MDC, multiple descriptions of the content are generated and sent over different channels to the receiver. If all descriptions are received, the decoder can reconstruct the source medium data stream using all of them to achieve minimum distortion. If only some of the descriptions are received, the decoder can still reconstruct the source medium data stream, however, with higher level of distortion. The reconstruction quality is proportional to the number of descriptions received. The more descriptions the decoder can receive, the less distortion it achieves. However, if the decoder receives a partial description or it cannot receive all bits of the description correctly in time, that description may be rendered useless. With scalable coding, a bitstream is partially decodable at a given bit-rate. With FGS coding, it is partially decodable at any bit-rate within a bit-rate range to reconstruct the medium signal with the optimized quality at that bit-rate. In most applications, FGS is more efficient compared with simulcast, and it is one of the most efficient schemes with continuous scalability.

8.2.3.3 Scalable Secure Media Streaming   To ensure access control to premium content, cryptography tools shall be used. The content provider stores an encrypted copy on the server. Upon receiving access requests, the content provider sends the encrypted medium stream, at the rate suitable for the network channel condition and receiver device capability, to the client devices. The questions is, when content encryption is needed, can we still provide the same level of scalability for fully or partially encrypted multimedia content for the same application? In other words, can my DTV, PC, PDA, and cellular phone all receive the same video with the quality adapted to the device capability as well as the network condition? Several schemes haves been proposed in the literature.

Stream ciphers (SC), which convert plaintext to ciphertext 1 bit at a time [2], offer means to decrypt an earlier portion of the ciphertext without the availability of the later portion of it; so does cipher block chaining (CBC) [2]. Therefore, the most intuitive way to provide scalable distribution of uncompressed and some types of compressed media using a single encrypted bitstream is to prioritize data, bit-by-bit or block-by-block, and encrypt the bitstream using SC or CBC [18].

Advanced scalable streaming media encryption schemes have been proposed in recent years. In Ref. 19, Wee and Apostolopoulos proposed a scalable streaming media encryption scheme (SSS) to enable transcoding without decryption. It utilizes CBC or SC to achieve progressive decryption ability. A fine-grained scalable (FGS) streaming media encryption scheme compliant to FGS coded video is presented in Ref. 18. In [20], Zhu et al. proposed two encryption algorithms for MPEG-4 that preserve MPEG-4's adaptation capability to varying network bandwidths and different application needs and enables intermediate stages to process encrypted data directly without decryption. Both schemes fully sustain the original fine-grained scalability of MPEG-4 FGS in the encrypted stream. Interested readers can see Refs. 18–20 for more details. Overall, the ultimate goal in designing a suitable streaming video encryption scheme for consumer networking is to offer consumers no-interrupted continuous playback, random access capability, and loss-resilient user experiences while providing adequate security. Using any of the aforementioned schemes, the server may send a single encrypted content stream to the client home, through the home gateway and may be a home media server, too, reaching each and every device desired to receive the content stream. The video decryption and playback are then performed on all devices with the quality adapted to the specific device capability and network condition. From a consumer's perspective, this provided scalability to various multimedia content, different network topology, changing bandwidth, and diverse device capabilities is valuable in offering satisfactory end-user experiences.

8.2.4 Alternative Approach for Preserving Content Copyright Without Sacrificing Consumer Convenience and Freedom of Use

Open Mobile Alliance (OMA) [21] DRM standard defines industry-wide interoperable mechanisms for developing applications and services that are deployed over wireless networks. To provide high interoperability, content and rights are logically separate entities, although being uniquely associated. The media object is first converted into DRM Content Format (DCF) that includes symmetric encryption of the content, either as a single object agnostic of the content's internal structure and layout for Discrete Media such as ring tones and images, or packet-by-packet for Continuous Media, making it useless to parties not having access to the Content Encryption Key (CEK). That is to say, the encrypted media object by itself, without the associated rights, is not usable on a device. This provides better opportunity for incentive-based schemes and other system solutions to help balance content protection with usage convenience.

The idea of incentive-based approaches comes from the following observation. In a content protection system, when cryptography algorithms are used, the keys can be the "weakest" point in the system. Once a key is known by the customer, the content can be easily given away or shared for free by a large group of people through the Internet. If economic incentives are given to a customer to encourage royalty-based sharing, not only shall it reduce content providers' loss due to free sharing, but also it will increase content providers' profit via peer-to-peer advertisement. To give an example, an incentive-based scheme that is built upon the super-distribution model defined in OMA is briefly discussed hereafter.

Figure 8.4 illustrates an incentive-based approach flow graph. Once Customer 1 purchased the content, the content (protected) is sent to the customer along with a ticket (ticket1) and the associated rights (①) The ticket is customer and content associated. Customer 1 later may send the protected content package along with her ticket (ticket1) to Customer 2 (②) for "sharing." This allows Customer 1 to share the information of the content without sharing the actual content freely and offers Customer 1 the opportunity to get an incentive rebate when her friend purchases the content. On the other hand, Customer 2 may obtain some content-related metadata or even a brief preview of the content from the packages she received from Customer 1. The ticket (ticket1) she got from Customer 1 may contain a promotional coupon or e-flyer that enables her to pay less for the same content. If she decides to have (buy) the content, she simply sends the ticket (ticket1) to the content provider along with the payment (③); she will receive the rights object for accessing the protected content as well as her ticket (ticket2 ④). Upon receiving Customer 2's payment, the server sends the incentive rebate to Customer 1 (④). Assume Cp is the protected content while P denotes the plaintext content; Tk, Rt, and U represent the ticket, the rights object, and the user, respectively. We have

$$P = f_{cp}(Cp, Rt);$$
$$Tk = f_{tk}(P, U).$$

That is, each ticket is associated with the content as well as with the customer. A protected content may be played back/used when the associated rights object is available. The customer can get the incentive rebate when the associated ticket is "used" by another customer. The more tickets she sends out, the more rebate she may get. This encourages royalty-based sharing and hence potentially increases a content provider's profit. Because the ticket may introduce coupon or reduced price to the shared party, it saves money for those using the ticket to purchase the content. Thus, it encourages the use of the ticket for content purchasing.

Clearly, the introduction of the incentive program does not affect the original capability, such as the interoperability, renewability, and content access scalability, of the system. For instance, if scalable encryption, such as those discussed earlier, is used, even if Customer 1 and Customer 2's device capabilities may be largely different,

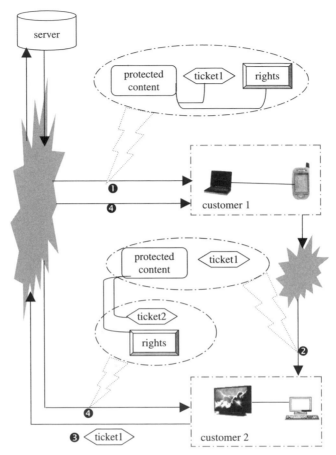

FIGURE 8.4 Encourage content distribution incentively without compromising the security of the content.

each will still be able to play back the content with the quality adapted to its device capability. The system is also scalable to the size of "sharing" and the incentive program model. It doesn't matter how many friends Customer 1 is sending her ticket to; the system security and the incentive program will not be affected. The content provider can also freely design its incentive program and change it at any time.

In the consumer network environment, after purchasing the content, Customer 1 can easily access the content within her home network using any devices because both the protected content and its rights object are accessible by any device on her home network. Because the incentive program discourages the customer to share her purchased content freely, others can hardly "share" the content without breaking into Customer 1's home network. The content is bounded by the home network or consumer domain, thus providing certain level of localization limitation.

Overall, the above proposed scheme encourages content distribution incentively without compromising the security of the content in an attempt to ease the conflict between content protection and user convenience.

8.3 PROVIDING USER-CENTRIC SERVICES FOR CONTENT PROTECTION IN CONSUMER NETWORKS

Until the day comes where all the intellectual property is free to use, content protection and conditional access mechanisms will be expected by the content providers in a highly connected consumer networking environment. However, in a dynamically evolving and highly integrated environment, simply relying on technology advances and licensing and legislation enforcement will not completely solve the problem and deter the situation.

The techniques discussed in Section 8.2.3 and Section 8.2.4 leverage customer interests with content protection. We call these user-centric designs. Customer-centricity (or user-centricity) is a prevailing measure of business success, especially evident in the consumer networked environment [1]. Low-price PC, digital camera, digital camcorder, MP3 player, CD-R, DVD-R, and various multimedia software tools made personalized multimedia creation and consumption easier than ever. Personalized broadcast multimedia consumption at home was also introduced via "set-top box" such as TiVo, ReplyTV, and DVRs. This trend of customization also influences media security requirements. Content protection specifications must be made more customer friendly by tying it to the media or a specific consumer adaptively with the usage of a combination of conventional and new security methodologies. This involves the development of a suite of tools that enforces access control without sacrificing user convenience. Together with licensing enforcement, one can expect to see new and better business models that make home entertainment and information sharing a better and better experience.

Marlin, developed by the Marlin Developer Community [16, 22] which was established in 2005 by Sony, Samsung, Philips, Panasonic, and Intertrust, is the first industry effort that aims at interoperability and consumer convenience within the home network environment. It adopts a user-centric approach to digital rights management and content sharing that targets licenses to users rather than devices. Founded on Intertrust's Octopus and Networked Environment for Media Orchestration (NEMO) technologies, Marlin is a DRM-based content sharing platform for consumer devices and multimedia services that is based on open standards and supports flexible consumer usage scenarios and business models.

Octopus is a general-purpose DRM architecture that can be applied to a variety of applications. It separates the protection of the content from the governance of that content to allow the issuance of rights to access content separately from information that governs where or when it can be used. This can be used to achieve greater end-user flexibility. For instance, a service provider can issue rights to a customer to use content but can allow that customer to independently manage which devices she would like to use that content on at any particular moment. By itself, Octopus is entirely content-agnostic, platform agnostic, and semantically agnostic. Marlin adopts Octopus technology for audio and video distribution in consumer domain.

The NEMO framework combines Simple Object Access Protocol (SOAP) Web services with Security Assertion Markup Language (SAML) authorizations to provide end-to-end message integrity and confidentiality protection, entity authentication, and role-based service authorization. Through the use of the NEMO framework, Marlin components can leverage a consistent mechanism to ensure that messages are delivered with appropriate protection and are exchanged between entities that are properly authenticated and authorized.

Marlin provides consumers of licensed multimedia content with the ability to share this content seamlessly across different devices and services in their home network or consumer domain. It offers technologies that allow users to acquire content through multiple distribution channels and to access it on any device that is part of their home domain. Its architecture [22] consists a set of specifications, such as the Marlin Broadband Delivery System Specification, the Marlin Broadcast Delivery System Specification, the Marlin OMA v.2.0 Gateway System Specification, the Marlin Physical Media Delivery Specification, and the Marlin Common Domain Specification, building on top of the Marlin Core System for developing a DRM-based digital media distribution system.

Will Marlin be successful at providing balanced digital multimedia business opportunity and "copyright protection" with "fair use?" Enforcing content copyright and access rights via improved user-centric services and versatile business model, together with new or enhanced content protection systems, perhaps can be expected to differentiate accessibility between legitimate and illegitimate uses in a home networked environment. This can be anticipated to benefit all parties in the content distribution chain and make user-centric business and service a reality.

REFERENCES

1. H. Yu, "Digital multimedia at home and content rights management." *Proceedings of IWNA4*, IEEE, January 2002.

2. B. Schneier, *Applied Cryptography*. John Wiley & Sons, New York, 1996.

3. T. B. Maples and G. A. Spanos, "Performance study of a selective encryption scheme for the security of networked, real time video." *Proceedings of 4th Inter. Conf. on Computer Comm. and Networks*, IEEE, 1995.

4. B. Furht, D. Socek, and A. Eskicioglu, "Fundamentals of multimedia encryption techniques." In *Multimedia Security Handbook*, CRC Press, Boca Raton, 2005.

5. I. Cox, et al., *Digital Watermarking*. Morgan Kaufmann Publishers, San Francisco, 2001.

6. J. Fridrich, "Image watermark for tamper detection." *Proceedings of IEEE ICIP'98*, 1998.

7. I. Cox, et al., "Secure spread spectrum watermark for multimedia." *IEEE Trans. Image Processing*, Vol. 6, pp. 1673–1687, 1997.

8. J. Fridrich, "Images with self-correcting capabilities." *Proceedings of IEEE ICIP'99*, 1999.

9. DVDCCA, Content Scramble System. Available at: http://www.dvdcca.org/css/.

10. 4C Entity, CPRM, CPPM. Available at: http://www.4centity.com/tech/cprm/.

11. MPAA, Broadcast Flag, Frequently Asked Questions. Available at: http://www.mpaa.org/broadcast_flag_qa.asp.

12. 5C, Digital Transmission Content Protection White Paper. Available at: http://www.dtcp.com/data/wp_spec.pdf.

13. Digital Content Protection LLC, High-Bandwidth Digital Content Protection. Available at: http://www.digital-cp.com/home.

14. AACS LA, AACS. Available at: http://www.aacsla.com/home.

15. 4C Entity, Content Protection System Architecture. Available at: http://www.4centity.com/data/tech/cpsa/cpsa081.pdf.

16. Marlin Developer Community, Marlin Developer Community Overview. Available at: http://www.marlin-community.com/images/wp/MarlinOverviewPaper.pdf.

17. DTLA, DTCP Volume 1 Supplement E Mapping DTCP to IP (Informational Version). Available at: http://www.dtcp.com/data/info%2020050228%20dtcp%20VISE%201p1.pdf#search='DTLA%20RTT'.

18. H. Yu, "Streaming media encryption." In *Multimedia Security Handbook*. CRC Press, Boca Raton, 2005.

19. S. J. Wee, and J. G. Apostolopoulos, "Secure scalable streaming enabling transcoding without decryption." *Proceedings of IEEE Int. Conf. Image Processing*, IEEE, Oct. 2001.

20. B. B. Zhu, C. Yuan, Y. Wang, and S. Li, "Scalable protection for MPEG-4 fine granularity scalability." *IEEE Transactions on Multimedia*, Vol. 7, No. 2, 2005.

21. Open Mobile Alliance™, OMA DRM Specification V 2.0, OMA-DRM-DRM-v2_0. Available at: http://www. openmobilealliance.org/.

22. Marlin Developer Community, Marlin Architecuture Overview. Available at: http://www.marlin-community.com/images/wp/MarlinArchitectureOverview.pdf.

9

DEVICE AND SERVICE DISCOVERY IN HOME NETWORKS

PAUL WISNER, FRANKLIN REYNOLDS, LINDA KÄLLSTRÖM, SANNA SUORANTA, TOMMI MIKKONEN, AND JUSSI SAARINEN

When a user brings a new device home, the device should be able to automatically integrate itself into the home network. Discovery protocols are the mechanisms that make this possible. This chapter addresses the question of how to make devices that integrate well and thrive in the home environment. Our particular focus is on discovery of devices and their services. What are the important technologies to understand? What are the basic operating principles? What new research remains to be done?

Home networks have become a common fixture in many homes. Recently, home networks have moved beyond the common case of sharing an Internet connection into entertainment and media applications. Some homes use their networks for entertainment, lighting control, and security. This has given rise to a new breed of home electronics: devices that utilize and participate in the home network. In this new world, devices compete to provide benefits to their users. The survival of new devices takes on a Darwinian flavor. The survivors will be those that provide the greatest utility, require the least of the user, and complement other devices. To achieve this end, devices must be able to automatically integrate themselves into the home network in useful ways.

We provide an overview of important or interesting discovery protocols that are most relevant to home networking of consumer products. We discuss the rationale for incorporating discovery protocols technologies.

Section 9.1 introduces the device and service discovery concepts and describes the common attributes of various discovery protocols. We list the discovery protocols that

Technologies for Home Networking. Edited by Sudhir Dixit and Ramjee Prasad
Copyright © 2008 John Wiley & Sons, Inc.

are most appropriate for home networks. Section 9.2 discusses the home environment and how discovery protocols are used in home networks. Section 9.3 addresses the issue of user control devices. How does the user exercise control over networked devices. Smart phones and PDAs have become excellent choices for control devices. Section 9.4 goes into the details of selected discovery protocols and discusses how these details are relevant in home networks. Section 9.5 describes areas in which researchers are attempting improve discovery protocols.

9.1   DEVICE AND SERVICE DISCOVERY

Discovery protocols are network protocols used to discover services, devices, or other networked resources. The ability to discover networked resources at runtime makes it possible to dynamically configure distributed systems.

Over the past 20 years, dozens of discovery protocols have been developed. Despite many years of practical experience with discovery protocols, it remains an active area of research for many organizations investigating topics related to scalability or security or context awareness.

Usually, a discovery protocol allows a service to be discovered on the basis of its type, its Application Programming Interfaces (APIs), and other properties—not just its name. For example, DNS (Distributed Name Service) is a name service. It resolves domain names to Internet Protocol (IP) addresses. For example, SLP (Service Location Protocol) clients can ask for services that match certain constraints, and servers respond with the names of services that match those constraints.

A *service* is an interface to a set of features. The service interface has methods and state. A *device* has a unique physical network address, for example, a unique Media Access Control Network Sublayer (MAC) address. A device may host one or more services.

In addition to physical devices, there are logical devices. A logical device is a software program running on a computer. At the IP protocol layer, a logical device may be indistinguishable from a physical device. The distinction has very little significance to the functioning of the system, but it is important to be aware that a device may be implemented in software on a computer.

9.1.1   Common Attributes

In general, there are four basic mechanisms that discovery protocols use for "discovery."

1. Advertisement: When a device joins a network, it sends a multicast message to "announce" its presence. For most protocols, these announcements are repeated periodically.
2. Inquiry: When a device wants to find other devices, it sends a multicast "inquiry" message. The inquiry message may contain details about the types of devices it is looking for. Devices receiving the inquiry respond directly to the inquiring device.

3. Directories: Directories are catalogs of available devices. When a device joins a network, it usually registers itself with a central directory. An inquiring device can then use the central directory to find devices it is seeking.

4. Description: Each protocol has a representation language to define the vocabulary and syntax used to describe the service and its properties. There is considerable variety in representation and query languages.

It is possible to forego the inquiry mechanism and discover devices by only listening to advertisements but it can take a long time to get advertisements from all devices. Queries provide a way to ask what other devices are present and to get responses right away. The general scheme is that, when a device first joins a network, it establishes an initial list of discoveries by performing a query. After that, it listens for advertisements to keep its list of discoveries up-to-date. Later we will see how the different protocols employ various additional mechanisms to track when devices have left the network.

Considerable variety can be found in the designs of different query protocols. Some protocols are optimized to minimize the size and number of packets required. Some are intended for simple, ad hoc networks, and others are intended for large-scale enterprises or even the public Internet. Many protocols have service registries that collect advertisements. It is much faster to query a central service registry, but it is hard to keep the registry up-to-date in very dynamic networks. A few protocols favor a fully decentralized, peer-to-peer approach to queries. Some protocols strictly limit the types of queries that are possible, whereas other protocols provide flexible pattern matching languages that allows for a wide range of possible queries.

Some discovery protocols also include "Description." Description provides a way for a device to describe its services, attributes, events, and methods. Each protocol has defined its own description language. Some description languages are extremely limited, and others offer a flexible syntax, rich vocabulary, and user extensibility. Description gives the advantage of feature extensibility.

On the other hand, some discovery protocols such as Bonjour and Bluetooth SDP do not provide descriptions of service interfaces. Here description is limited to lists of properties related to the particular service. The interfaces of predefined service types are known to the programmer of the client program, so there is no need for the client program to examine a description to determine what methods and attributes are available.

9.1.2 Interoperability

The wide range of protocols, representation languages, and query languages has resulted in very little, if any, interoperability between the various discovery protocols. As a consequence, most commercially available systems implement multiple discovery protocols.

Devices often implement several discovery protocols so that they can operate in different types of environments. The most complex case is the device that uses multiple discovery protocols simultaneously. The complexity is often hidden from the user behind a graphical user interface.

Supporting multiple protocols has major drawbacks. Many mobile devices are memory constrained or have inadequate user interface capabilities. This limitation makes it very undesirable to use multiple protocols when, at least theoretically, a single protocol would be sufficient. Users of laptops or desktops are not usually concerned with the cost or power consumption of the network they are connected to. Mobile phone users may be concerned inefficient protocols designed for Local Area Networks (LANs)—both the financial cost of using the cellular network and the drain on their batteries of using any network.

If interoperability with existing systems is not a goal, then one could create a new proprietary discovery protocol technology. This is a common choice for very specialized systems and for researchers developing new kinds of discovery protocols. But developers of consumer products should consider the need to interoperate with other devices using the standardized protocols.

9.1.3 Distributed Middleware Toolkits

It is common for discovery protocols to come as an integrated part of a distributed middleware toolkit. In addition to discovery, distributed middleware toolkits provide for remote invocation and events. For example, SSDP (Simple Service Discovery Protocol) is part of UPnP (Universal Plug and Play); a suite of distributed computing technologies that includes SSDP, SOAP (Simple Object Access Protocol), and GENA (Generalized Event Notification Architecture). Likewise, Jini is a distributed middleware toolkit that provides its own services discovery protocol.

The template of Figure 9.1 will be used to illustrate the correlation of features of different protocols. For each protocol, we will show how it maps onto the template. One advantage of a discovery protocol being packaged with a distributed middleware toolkit is that the discovery protocol has already been carefully designed to work well with the other components of the system. The disadvantage of such integration is that it can be difficult to use a mix of features from different toolkits.

The discovery protocols that are most relevant to the home are

- SLP (Service Location Protocol) [1]
- Bonjour [2]

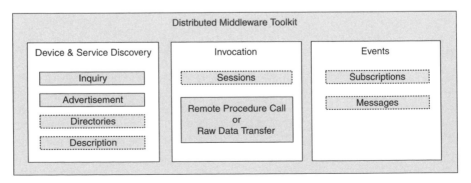

FIGURE 9.1   Template of typical features of distributed middleware toolkits.

- UPnP / SSDP—Simple Service Discovery Protocol [3]
- Jini [4]
- JXTA [5, 6]
- DHCP [7]
- Bluetooth SDP [8]
- Web Services Dynamic Discovery [9]
- eXtensible Service Discovery Framework [10]

A common prerequisite to discovery is that devices already have IP network connectivity. Bluetooth SDP is an exception because it does not use IP at all.

9.1.4 Other Discovery Protocols

There is a much larger list of discovery protocols that are not intended for use in home networking. Yet, a researcher interested in pushing the frontiers may want to study them for ideas that could be applied to the home.

There are many other different IP-based discovery protocols. Some examples include the following: MIT Intentional Names [11], UCB Ninja [12], FreeNet [13], Cooltown [14], Salutation [15], SyncML [16], and Sony FEEL.

9.1.5 Directory Services and Other Configuration Management Systems

Directory Services such as X.500 [17], LDAP [18], JNDI, UDDI [19], Novel NDS [20], Banyan Vines [21], CORBA [22], DCE [23], and DCOM [22] provide an entirely different mechanism for finding network resources. Directory Services are distinct from discovery protocols because they require an administrator to make entries into a directory for each service.

At a sufficiently high level of abstraction, lots of configuration management systems have similarities to discovery protocols. Dynamic shared libraries, hot-swap technologies like "plug-and-play" or PCMCIA card managers and even Wireless Access Protocol (WAP) EFI (External Functional Interface) provide a type of late binding or name resolution functionality. SIP [24], USB negotiation, CC/PP [20], SyncML, and Web search engines all provide limited mechanisms for describing, advertising, and discovering some sort of resource.

Some protocols involved in establishing the initial IP connections employ a kind of discovery mechanism. Wireless 802.11 LAN devices advertise themselves by emitting a beacon message 10 times per second. DNS is used to find a device's IP address based on a network name. The 802.11 beacons and DNS are critical to automating network configuration, but when we write about Discovery Protocols, we are referring to protocols that establish connections between applications and services.

Although there is some merit to drawing parallels between these systems and discovery protocols, none of these services were intended to serve as a general purpose discovery protocol. Google, for example, is more like a "distributed grep" than a discovery protocol.

9.2   THE HOME AND THE EXTENDED HOME

In the previous section, we described the discovery protocols that are most relevant to the home and talked about their common attributes. In this section, we will describe the special characteristics of the home that are relevant to discovery protocols.

Typical homes tend to have quite simple networks. Usually, these networks have a single router used primarily to share access to the Internet. In such networks, most devices will exist on the same local area network. The set of available devices and services is dynamic because many devices will be powered-on only when they are needed.

Consumers are reluctant to make significant investments in network infrastructure—so products must be able to function in home networks with little or no dependency on network infrastructure services. The components of home networks and home networking applications tend to be developed by independent third parties and incrementally deployed by consumers. Consumers must manage their home networks themselves.

For these and other reasons, deployment and management of home networks must be as simple as possible. Technologies, such as service discovery, that enable auto-configuration are essential to widespread success of home networks.

9.2.1   Characteristics of the Home Environment

Developers can probably assume that the multicast messages used for device discovery will be able to reach every device. A wireless home network is usually 802.11g but might also be 802.11b, 802.11a, or 802.11n. Bluetooth devices may also be present. While it is possible for Bluetooth devices to implement IP over Bluetooth (Bluetooth PAN [25]), it is uncommon. It is safer to use the native Bluetooth protocols such as Object Exchange (OBEX) or Radio Frequency Communication (RFCOM).

Today, wireless networks have insufficient capacity for reliable delivery of high-bandwidth streaming media. The theoretical peak capacity is sufficient—and standard definition video does work well in optimal condition. Unfortunately, outside the lab the realities of the radio layer decimate the optimum. Video playback stutters or stops. Actual wireless bandwidth is a function of the total use of the radio frequency range of the network. 2.4 GHz wireless bandwidth is reduced by interference from cordless phones, Bluetooth, and especially other nearby wireless networks.

One remedy is to use wired network connections—in this case, the connections usually provide more capacity and are more reliable. Media can also be delivered by a video cable or some other transport besides the network The upcoming 802.11n and Multiple-Input, Multiple-Output communication (MIMO) endeavor to quadruple the bandwidth of 802.11g with particular efficiency for streaming media. We will have to wait before we will know whether 802.11n truly solves the problems.

Home devices are usually network peers because there is no central infrastructure or server that organizes or facilitates the communication. One cannot assume the existence of network infrastructure elements such as DHCP servers, DNS servers, SLP servers, UDDI servers, and so forth. Instead, the end-points discover and communicate with each other directly. The applications include media and file sharing, multiplayer games, entertainment applications, and local "chat."

9.2.2 Characteristics of the Extended Home Environment

The extended home concept involves the Internet and visitors.

Some home devices make requests of Internet-based services using wide-area remote procedure calling protocols such as XML_RPC and SOAP. Two examples are Digital Video Recorders that access television program guides and MP3 players that access CD information lookup services.

Remote access is another potential feature of the extended home. This remote access allows you to control devices in your home from anywhere on the Internet. A network may include remote devices connected by a virtual overlay network (VON) such as a virtual private network (VPN). The network topology of a VON can make distant devices appear to be logically local because it is within a small number of network hops in the virtual network. It is common for service discovery protocols to assume that a device is nearby if it is logically local.

In both the home and the extended home, you should consider visitor scenarios. What level of access should a guest device have? How easy should it be for the visitor to obtain access? It is desirable to make it possible for mobile visitors to automatically, or at least very easily, connect to and disconnect from a home network. The owner of the home must somehow specify which devices visitors are authorized to use.

Years from now, an additional complication may arise: it may become common for visitors to be wearing their own private personal area network in their clothing. These mobile networks would have a mobile device as the "hub." All the devices connected by this type of network are owned and administered by the visitor. Personal mobile networks may include cameras, microphones, speakers, storage systems, GPS receivers, personal printers, smart toys, smart sports gadgets, and so forth. What interactions should be possible between the visitor's personal network and the home?

Our discussion of visitor scenarios evokes many questions, but we have very few answers. There are many research opportunities involving visitors.

9.3 USER CONTROL DEVICES

What is the best way for users to control their devices? Today, we accept that device controllers are spread around the home and each device has its own dedicated controller. But standardized protocols make it possible to have a unified control device. What device should host the user interface to the smart home?

The most common controller today (2006) is still the infrared remote. They are found piled on coffee tables and hiding under couch cushions. The infrared remote is a dumb controller that (1) controls a stationary device, (2) has no

TABLE 9.1   Client Devices and Capabilities

	Installable Third-Party Software	Support Java	Support Java Mobile Code (Class Loader)	Wireless LAN	Bluetooth	Bluetooth PAN (TCP/IP)
Laptop computer	Yes	Yes	Yes	Yes	Yes	Yes
PDA	Yes	Yes	Yes	Yes	Yes	Some
Mobile smart phone	Yes	Yes	2006	2005	Yes	Some
Common mobile phone	Java only	Limited	No	No	Yes, restricted	No

awareness of the state of the device, and (3) does not determine whether a control message was received.

Distributed middleware toolkits make it possible to create smart controllers that (1) use discovery protocols to integrate themselves with the home devices, (2) are aware of the device state and display information from the device to the user, (3) receive responses from control messages, (4) can authenticate themselves as authorized for the device, and (5) coordinate the actions of many devices.

Handheld computers such as PDAs, Mobile Smart Phones, and laptop computers can serve as smart client devices from which the user can control and otherwise consume services. Table 9.1 shows some of the key devices and their capabilities.

Handheld computers such a smart phones make excellent client devices because, unlike dumb infrared emitters, they can receive feedback from the devices they control. The feedback might be an update to the device state or information retrieved from the device. For example, a digital media controller can retrieve a list of songs available on a media server and display the list on the screen of the mobile computer. A print controller can update the screen when a print job has been interrupted by someone opening the paper tray. We have found that discovery protocols work well even on devices with the most limited networking capabilities such as cell phones running Bluetooth PAN. In 2005/2006, manufacturers were introducing many mobile phone models with WLAN capability.

Mobile phones and PDAs can be programmed in C, C++, Java, or Python. Usually, C and C++ allow greater access to device features such as networking, camera, and flash storage. Devices programmable in C and C++ include those based on Symbian OS, Microsoft Windows Mobile OS, Palm OS, and Linux OS. Symbian OS phones are selling at twice the rate of all others combined. C and C++ are used to create software that uses advanced device features or requires high performance. Unfortunately, C and C++ are normally not compatible across different devices or operating systems.

Java for mobile devices (J2ME) has restrictions designed to protect your device from malicious software. For example, there is no access to the file system and a J2ME program cannot initiate a phone call without prompting the user. These restrictions are a security measure that allows you to download an untrusted application without having to worry that it can damage your data. Another cause of restrictions is the desire to standardize interfaces: because we want interfaces to be standardized across all Java devices, they are not made available until a lengthy standardization process has concluded. Java application development is much faster than C++ because it is easier to program and easier to debug. Java programs work across different devices from different vendors running different operating systems. If your goal is to create software that will run on the largest number of mobile phones, then Java is the best choice.

Python is available for Nokia S60. Python is easy to learn and fast to program. The Python community is expanding the library of functions very quickly. It is relatively easy, given knowledge of C++, to extend the library yourself with C++. Python is an excellent choice for rapid prototyping.

9.4 SELECTED DISCOVERY PROTOCOLS

We have selected a list of 10 discovery protocols that should be considered for home networks. These protocols are Bluetooth SDP, UpnP, SSDP, SLP, Jini, JXTA (JuXTApose), Bonjour, DHCP (Dynamic Host Configuration Protocol), WS (Web Services)-Discovery, and XSDF (eXtensible Service Discovery Framework). Clearly, service discovery remains an active topic of research in that there is quite a range of discovery protocols in current use.

Different protocols are specialized for different applications. And yet, though each discovery protocol is different, the fundamental problem solved by each service is similar.

Given this similarity, it is not surprising that study of different discovery protocols reveals common issues, design patterns, or components shared by most if not all of the services. We have found it useful to characterize discovery protocols by their approach to the following issues:

- Advertisement: How do device and services make their presence known?
- Inquiry mechanism: How do clients find devices and services.
- Service registry: Are there service registries? How they are organized?
- Description: How do devices and services describe themselves?

In the following sections of this chapter, we discuss several different discovery protocols. The coverage is restricted to protocols that are appropriate for the extended home.

9.4.1 SLP

SLP is widely used in production environments but also as a base protocol for many research projects. In SLP, each service has a type that consists of an abstract type, used naming authority, and a concrete type. The naming authority may be omitted if it is Internet Assigned Numbers Authority (IANA). The abstract type of a service describes it in a very general manner, and the concrete type tells for example what protocol can be used to access the service. In addition, services can have attributes as attribute strings.

SLP is a flexible protocol because it supports both multicast discovery, for simple, self-organizing networks, and centralized registry–based discovery, for larger, enterprise-scale networks. Queries can be transmitted using synchronous or asynchronous messages. Using the SLP directory agent discovery protocol, clients can automatically determine when to use peer-to-peer queries or registry/directory-based queries. In either case, successful query replies can return either the URI of the matching object or matching description. Figure 9.2 shows the SLP in the distributed middleware template.

The SLP architecture defines User Agents (UA), Service Agents (SA), and Directory Agents (DA). User Agents present control interfaces to a user, send queries, and collect replies. Service Agents are associated with services and respond to queries. Directory Agents provide catalogs of available services. In addition to DAs, SLP also provides direct P2P discovery via multicast methods.

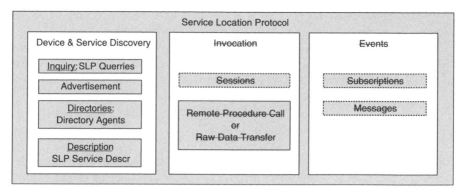

FIGURE 9.2   SLP in the distributed middleware toolkit template.

Advertisements and queries can be tagged with administrative scopes. Scopes can be used to differentiate between directories allowing advertisements and queries to be efficiently routed to appropriate directories.

Unlike current DNS, SSDP, and many other simple Discovery protocols, SLP includes an authentication protocol to protect against unauthorized modifications to a directory. Though useful, it is worth noting that this is not a complete security story.

SLP uses relatively simple and compact description and query languages. A service description has a simple data model based on a list of name/value pairs. IANA is the governing body that controls the specification of service descriptions and the vocabulary of terms that can be used in an SLP description.

The following listing is a simple example of an SLP Service Description:

```
#Register a testing service
service:test.openslp://192.168.100.1,en,65535
copes=test1,test2
description=Testing Service
title=Engineer
version=1.0
authors=Reynolds,        Racz, Hietaniemi
```

The SLP query language is a simple and elegant pattern matching language. Wildcard patterns, AND, OR, and NOT operations are supported.

A few examples of SLP queries:

```
(authors=Reynolds)
(!(authors=Reynolds))
(&(title=Engineer)((description=Test*)(authors=R*)))
```

SLP has been deployed by Novell, Sun, and other companies—but not Microsoft. Internet Engineering Task Force (IETF) and 3GPP working groups are considering making SLP part of the extended SIP specifications. It is worth noting that Apple

Computer was an early supporter of SLP, but they have been discouraging its use since the announcement of Apple's Bonjour discovery protocol.

There is an IETF experimental RFC of the SLP specification that describes a publish-and-subscribe service based on SLP advertisements and queries. It has the advantage of using UDP-based multicast protocols, which improves its performance, but it weakens its reliability. This is an experimental protocol and it is not in common usage.

9.4.2 Bonjour

Bonjour (originally named Rendezvous) is Apple's system for service discovery that uses Multicast Domain Name Server (MDNS), a link local multicast variant of DNS [26, 27], and Domain Name System (DNS) Service Record (SRV) [28] records (see Fig. 9.3). DNS SRV records provide terse descriptions of services available on a device.

Three examples of SRV records are

```
ftp.tcp        SRV 0 0 21 server.asdf.com.
finger.tcp     SRV 0 0 79 server.asdf.com.
 http.tcp.www SRV 0 0 80 server.asdf.com
```

The basic operation of Bonjour is simple. A client sends multicast queries to a special address and port number. Every device in multicast range responds giving its IP address. Next, the client attempts to retrieve the SRV records from each responding device. The SRV records describe the services on the device.

Microsoft has created their own variant of link-local multicast DNS, known as Link-local Multicast Name Resolution (LLMNR) [29]. LLMNR and mDNS are similar, but there are small differences. At the time this chapter was written, neither protocol had been standardized by the IETF—though LLMNR has been adopted by the Bluetooth SIG as part of their PAN [25] (IP over Bluetooth) profile.

Bonjour has some similarities to SSDP [3]. Its queries are limited to simple string matching. Unlike SLP, there is no support of complex pattern matching. The design of its protocol emphasizes peer-to-peer operation in ad hoc networks. A multicast

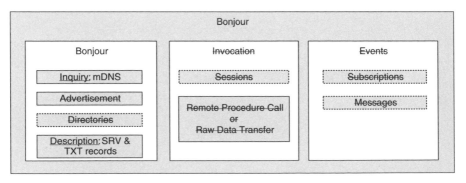

FIGURE 9.3   Bonjour in the distributed middleware toolkit template.

protocol is used to send announcements of services. Hosts can cache these announcements to reduce the need to send queries. Queries if needed are also multicast to all hosts. This removes the need for a registry of currently available services. Unlike most protocols, the replies to Bonjour queries are also multicast to all hosts. This allows all hosts to eavesdrop on the replies, and this information can also be used to populate a discovery cache.

Each service listed by DNS SRV may provide additional properties by supplying a DNS TXT document. DNS TXT can contain any text up to a maximum size of 64 kb. The DNS TXT format is described in Ref. 30. It is based on simple name/value pairs, such as "Color=4," "PaperSize=A4," or "InstalledPlugIns= JPEG,MPEG2,MPEG4." This is a simple and extensible approach that can be easily enhanced if necessary. Apple Computer has invited other venders to join them in the creation of standardized service descriptions.

There are several IETF working groups investigating approaches to service discovery. Of particular interest are about half a dozen proposals for enhancements to DNS including Dynamic DNS Updates, Trusted DNS management, Multicast DNS and DNS SRV, and RR records. DNS is already the name resolution protocol for the Web. Taken collectively, these new DNS features may eliminate the need for most other discovery protocols.

9.4.3 Universal Plug and Play/SSDP

The Simple Service Discovery Protocol (SSDP) [3] is a feature of the Universal Plug and Play (UPnP) distributed middleware toolkit, Figure 9.4.

UPnP is a service advertisement and discovery architecture supported by the UPnP forum headed by Microsoft. UPnP aims to standardize the protocols used by devices in XML-based communication. In addition to service advertisement and discovery, the UPnP specification also describes device addressing, device control, eventing, and presentation. Eventing allows clients to observe for changes in the discovered service. Presentation allows clients to obtain a generic user interface to a discovered service. Control is expressed as Simple Object Access Protocol (SOAP) objects and their URLs in the XML file. Control is established by sending SOAP messages to these objects.

The control point sends HTTP messages to the device that specify a method name and argument list. The device processes the message, performs the action, and optionally sends a response back to the control point. The response has a return status (success or failure) and may include an XML document containing further results.

SSDP is intended for P2P LANs—it is based on multicast queries and multicast advertisements. SSDP messages are formatted using HTTP-like headers and encoding rules. UDP-based versions of HTTP, HTTP/U, and HTTP/MU carry SSDP queries and replies. The protocol currently has no support for directories, though adding directories should be possible.

SSDP utilizes an XML-based description language. SSDP descriptions are structured and quite rich. They include much more information than found in the descriptions of most other discovery protocols. In addition to attributes usually associated

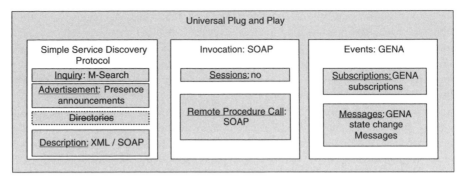

FIGURE 9.4   UPnP and the distributed middleware toolkit template.

with a service, a SSDP description will typically include pointers to programmatic descriptions and user interface descriptions. The XML description document includes four important items [31]:

- A presentation URL: provides a Web-based graphic user interface (GUI) for device control
- A control URL: the address to send device-specific commands to control the device
- An event subscription URL: the address used to subscribe to the device's event service
- A service control protocol (SCP) definition: describes methods that can be used to interact with the device

There is some discussion of UPnP converging with Web Services. This is because there are similarities and common technologies. For example, Web Services is also based on HTTP and XML and utilizes SOAP for remote invocation.

In UPnP, Control Point applications use SSDP to discover devices they would like to control. UPnP devices can be either physical devices, such as a printer, or they can be software programs running on a computer. UPnP devices contain one or more UPnP service. The actual control interaction is through the services.

SSDP provides two mechanisms to discover Devices. Normally, both mechanisms are used concurrently.

Mechanism No. 1: Presence Announcement**:** Programs can listen on a multicast channel for *ALIVE* messages. Each UPnP device or service multicasts its ALIVE messages immediately after start-up. In addition, each device or service periodically repeats its ALIVE advertisements. Presence Announcements are typically sent at 30-minute intervals, however implementers may choose a different interval.

Mechanisms No. 2: Discovery Inquiry**:** Programs can multicast an *M-Search*. This is a multicast message that advises each UPnP device to send (unicast) a presence announcement to the requesting control point.

The search criteria of an M-Search message is limited to one of:

* all devices and services
* all root devices
* devices with a given device UUID
* devices of a given type
* devices supporting a given service type

Both ALIVE messages and M-Search replies are UDP datagrams limited to 1500 bytes. UPnP devices announce when they are shutting down by multicasting a *ByeBye* message. These messages are similar to ALIVE messages. When a control point first joins a network, it sends a multicast M-Search message to establish an initial list of available devices. Afterwards, the Control Point will listen for Presence Announcements and ByeBye messages to learn about new Devices joining or leaving the network.

ALIVE messages and M-Search carry a URL that points to the device description document. A control point must retrieve the device description document before it can control a device. All description documents are in XML. The verbosity of XML contributes to the size of the descriptions but makes it easy to add structured data.

Unlike protocols such as SLP, SSDP has no query language; instead there are M-Search inquiry messages that specify one matching criteria.

Because there is no query language, all but the most simple of queries must be constructed from multiple queries as well as retrieval and evaluation of the descriptions.

In many ways, SSDP protocol is functionally inferior to SLP. SSDP does provide an abstract data model that is rich and structured, but it is quite verbose and costly to retrieve. It does not support directories, it does not provide a query language, and it uses an inefficient, HTTP-derived multicast protocol. SSDP has the advantage that it is simple to implement. The fact that Novell, an intense competitor of Microsoft, supports SLP may be another advantage for SSDP within Microsoft.

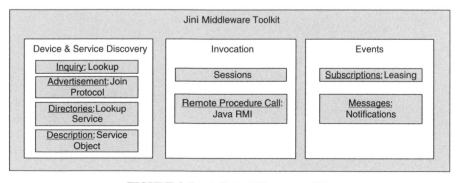

FIGURE 9.5 A Jini middleware toolkit.

9.4.4 Jini

Sun created the Jini protocol and contributed it to a standard body named the Jini Community (http://jini.org).

The Jini architecture is Sun Microsystems' attempt to realize their vision of making a distributed system consisting of many different devices connected with a network to form a system like one large computer. In other words, this means moving from disk-centric to network-centric thinking. Discussion in this section is based on Refs. 4 and 32.

Fundamentally, Jini is a distribution framework built on top of Java and Java/RMI (Remote Method Invocation) see Figure 9.5. The main advantage of Jini is the support for very dynamic and flexible distributed systems, in which participating devices (hosts) can join and leave without any need of administrative work. A Jini system consists of three main types of components: clients, service providers, and lookup services. The main components and operation of Jini is shown in Figure 9.6. Service providers implement and offer services. Service, which is the most important concept within the Jini architecture, is an entity that can be used by a person. This can be a computation, storage, communication channel, filter, hardware device, or even another user. For example, printing a document could be a service. They can form hierarchies of services by using other services. Services communicate with each other using service protocol, which typically (but not necessarily always) runs on top of Java/RMI.

Central to the Jini discovery process is the Lookup Service. A client wishing to find a service can query its lookup service for an appropriate match. Jini has a subtly different meaning for the word "discovery." In Jini, a Service Provider must "discover" the Lookup Service and then "join" the lookup service. Once the service has joined, it will be a potential result for future queries from clients.

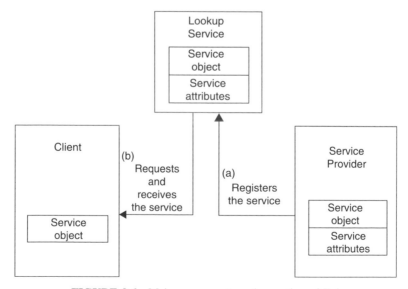

FIGURE 9.6   Main components and operation of Jini.

It seems obvious that the Lookup Service should live in the fixed network.

Lookup services act as proxies to which services are registered and from which they can be found by clients or other services. Lookup services work like yellow pages in a phone book, in which services are registered and found by their interface and possible additional descriptive properties. Service providers find lookup services with discovery protocol. Services are added to a lookup service with join protocol. Clients use a different protocol, lookup, when they are searching for services.

The actual service discovery functionality needed to find lookup services is based on multicast requests and announcements. Lookup services advertise themselves with announcements, and they can also be found with multicast requests. The idea behind having both is that when a new device is plugged into a Jini system, it first actively searches for lookup services with multicast requests. After a while, it ends the active search and switches to passive mode, only listening to possible advertisements of lookup services.

Despite their many functions, lookup services are not an absolute necessity in a Jini system. The Jini specifications include also a technique called peer lookup, which can be used to find and use services if no lookup services are available. In it, the client issues the same request, and service providers then register with the client as they normally would have done with the lookup services.

Peer lookup is implemented by putting a Lookup Service in each device that is running a Jini client. If every device has its own Lookup Service, then you have a peer-to-peer discovery process.

The main advantage of Jini is in its use of mobile code. Jini Services supply a Service Object to the client. A Service Object is a unit of Java byte code. The client retrieves the Service Object that is assumed to implement the Java interface for the particular service type. The client directly calls the interface of the Service Object. The Service Object uses Java RMI for communicating over the wire. It is transparent to the client whether the Service Object actually invokes remote services or implements services to execute locally.

We, the authors, have a great deal of affection for Jini and often lament that it has not been widely adopted for use in the world of mobile and embedded computing. We speculate that the reasons are (1) J2ME, "Java 2 for Mobile and Embedded," lacks the class loader and other features required by Jini, and (2) mobile device designers want well-known memory requirements. However in 2006, more extensive J2ME implementations are appearing on mobile devices. In particular, those that support Personal Profile are capable Jini client devices.

9.4.5 JXTA and JXTA Search

JXTA (JuXTApose) is a collection of protocols and XML-based tools for the creation of peer-to-peer applications (Fig. 9.7). JXTA is a Sun-sponsored, open source program that is marketed primarily as a research activity—rather than a "product." The basic framework consists of the core P2P protocols and five key abstractions: uniform peer ID addressing (based on UUIDs), peer groups, XML-based advertisements, resolvers (the discovery service), and pipes.

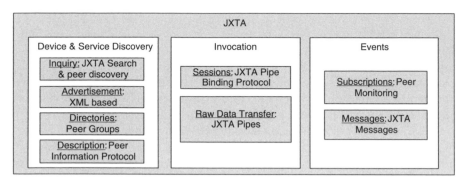

FIGURE 9.7   JXTA search in the distributed middleware toolkit template.

Each peer group can implement their own discovery service based on JXTA advertisements. JXTA resolvers provide both an API for peer group–specific discovery services and a simple, bootstrap discovery service. Higher-level search services (such as JXTA Search) are built on top of rendezvous peers.

JXTA Search is a peer-to-peer meta-search engine. Its architecture is based on a set of interconnected search hubs. JXTA hubs are used to aggregate information registered by providers. Provider registrations advertise a provider's metadata and structure, associated query space, matching predicates, and an XML document describing the types of information provided. Queries are formed using an XML-based query language, and these queries are routed to appropriate JXTA hubs. JXTA Search does not perform any presentation of provider responses. It collates results from providers, performs ranking on the results with respect to the query, and returns them to the requesting client. The client performs the final presentation of the results.

The JXTA Search framework consists of:

- *Query Routing Protocol (QRP)*: XML protocol for queries, responses, and registrations
- *Query Spaces*: define the structure of a query and its associated registration
- *Registration*: XML-based encoding of a logical statement characterized by a query space
- *Query Formulation*: XML-based queries adhering to specific query spaces
- *Query Resolution*: queries are resolved by a resolver
- *Query Routing*: queries are routed to the appropriate provider by sending XML requests over HTTP
- *Provider Responses*: providers respond to queries in arbitrary XML

Both JXTA and JXTA Search provide a framework for supporting multiple discovery protocols. JXTA resolvers provide a consistent API to service discovery. JXTA Search goes further. It provides mechanisms for translating between the meta-vocabulary and the local vocabularies associated with each "hub."

9.4.6 DHCP

The Dynamic Host Configuration Protocol [7] is a commonly deployed service that is used to configure Internet hosts (Fig. 9.8). The protocol has a simple client–server mode of operation. Clients use DHCP to acquire configuration information when initially connected to a network. The most important, or at least the most commonly used information provided by DHCP servers are IP addresses for clients, the address of the default gateway or router, and the addresses of the default DNS servers. Additional protocol control parameters like ARP cache timeout or the default value for TCP TTL can also be provided by DHCP.

DHCP also provides a rudimentary discovery protocol. The discovery protocol exists to find the services involved in configuring the basic network. There are 255 DHCP Option tags assigned by IANA [33]. Some of these tags are used to identify important network infrastructure services. In addition to routers and DNS servers, DHCP can be used to find services such as SMTP, LDAP, SLP, HTTP, SIP, NTP, NETBIOS [34], and so forth.

DHCP discovery is not generally used for configuring application-level services that are specific to a particular site, enterprise, or network. However, it can be used to discover other discovery services such as SLP directory agents or LDAP directories. For example, RFC 2610 describes, in detail, how DHCP can be used to discover SLP directory agents and configuration information intended for SLP user agents. Because of its role in bootstrapping and configuring network hosts, DHCP is a moderately complex protocol. Much of this complexity is associated with the allocation and management of IP addresses. The service discovery function of DHCP is fairly straightforward. In general, clients attempt to discover the existence of a server by broadcasting a link local DHCPDISCOVER message. If a server hears this message, it will reply with a DHCPOFFER message that includes an available IP address and a list of configuration parameters that can be obtained from the server. It is possible that multiple servers will receive the DHCPDISCOVER message. Each server will send its own reply.

The client then broadcasts a DHCPREQUEST message specifying a server. The other servers see the request and notice that the client is not interested in talking to

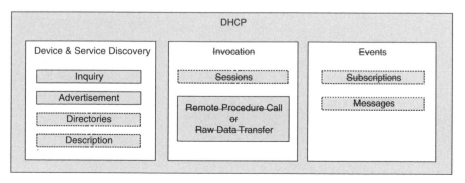

FIGURE 9.8 DHCP in the distributed middleware toolkit template.

them. This means they are free to offer their IP addresses to other hosts. The
DHCPREQUEST includes the client's list of desired configuration parameters,
such as default router, DNS servers, and so forth. The names of configuration parameters are integer values ranging from 0 to 255. The options field in the
DHCPREQUEST message is simply a list of 8-bit values that name the desired parameters. The server replies with a DHCPACK message that includes the configuration parameters.

Initially, clients discover and communicate with DHCP servers via link-local
broadcast. Once a client knows the IP address of a DHCP server, it can communicate
via unicast. Networks that have multiple link-local segments either need multiple
DHCP servers or they can use Relay Agents that relay link-local packets to a
remote DHCP server.

DHCP provides mechanisms for coordinating with DNS so that a host can preserve its DNS name when it is allocated a new IP address; however, unlike a discovery protocol like SLP, it does not provide a protocol for applications or hosts to add to
or modify the services that can be discovered using DHCP. This is not usually a
serious limitation as most application or site-specific services will be advertised
and discovered using other, higher-level discovery protocols.

9.4.7 Bluetooth SDP

Bluetooth Service Discovery Protocol (SDP) is a protocol designed to solve service
discovery problem between Bluetooth enabled devices [35], (Fig. 9.9). It does not
provide access to services, brokering of services, service advertisements, or service
registration, and there is no event notification either. Bluetooth SDP supports searching by service class, search by service attributes, and service browsing [36]. Methods
of invoking the found services are outside the scope of SDP [35]. The SDP API also
supports stop rules that limit the duration of searches or the number of devices
returned [31]. Devices are identified by Universally Unique Identifiers (UUIDs),
which are generated once at the time a service is generated. They are used to avoid
service definition collisions.

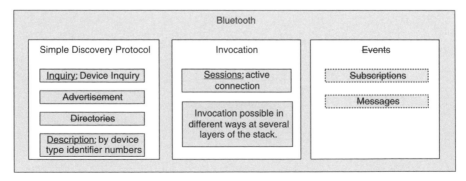

FIGURE 9.9 Bluetooth SDP in the distributed middleware toolkit template.

Bluetooth devices must find each other and establish connections before service discovery begins. Device discovery starts with an inquiry mode in which one device is in inquiry mode and the devices waiting to be discovered are in inquiry scanning mode. Once two devices have discovered each other, a connection is established using paging mode. The act of connecting adds a new device to a piconet. Bluetooth piconets are composed of a single master and up to seven slaves.

Bluetooth SDP is basically a client–server protocol. The server maintains a list of service records that describe the characteristics of services associated with the server. Each record contains information of a single service. The client can retrieve information from a service record by sending an SDP request. It has to open a separate connection to actually use the service as SDP does not provide any mechanisms to do this. One Bluetooth device may contain exactly one server. A single device can operate as a server and as a client at the same time.

SDP (Service Discovery Protocol) is a simple, Bluetooth-specific discovery protocol that is local to a single piconet. A Bluetooth piconet is a group of up to eight devices that are connected using Bluetooth. SDP is layered on top of Bluetooth protocols, including L2CAP. The description or "service class" of a Bluetooth service is a record consisting of a flat list of attributes or name/value pairs. A similar, relatively simple abstract data model is used by SLP and other discovery protocols. A Bluetooth service class is specified using a Bluetooth-specific description language and vocabulary.

A service record maintained by the SDP server consists entirely of a list of service attributes. Each service attribute describes a single characteristic of a service. Service attributes consist of two components: an attribute ID and an attribute value. Each service is an instance of a service class. The service class definition provides the definitions of all the service's attributes. A service record contains service attributes that are specific to a service class and some attributes that are common to all services. Service searching is based on UUIDs, and service browsing is based on attributes of the services [37].

A few examples:

IrDA-like Printer

```
(( L2CAP, PSM=RFCOMM ), ( RFCOMM, CN=1 ),
( PostscriptStream ))
```

IP Network Printing

```
(( L2CAP, PSM=RFCOMM ), ( RFCOMM, CN=2 ), ( PPP ),
( IP ), ( TCP ), ( IPP ))
```

Synchronization Protocol Descriptor Example

```
(( L2CAP, PSM=0x1001 ), ( RFCOMM, CN=1 ), ( Obex ),
( vCal ))
```

```
((L2CAP, PSM=0x1002), (RFCOMM, CN=1), (Obex),
(otherSynchronisationApplication))
```

Services advertise their services to the master of a piconet when the device hosting the service joins the piconet. Clients send queries to the piconet master that hosts the SDP service registry.

9.4.8   Web Services Dynamic Discovery

Web Services Dynamic Discovery (WS-Discovery [9]) is a multicast discovery protocol for locating Web services (Fig. 9.10). It allows discovery of services in ad hoc networks that have minimum of networking services (e.g., DNS or directory services).

The basic architecture of WS-Discovery has two entities: client and target service. The clients can search target service by type, scope, or both, by sending probe message to a multicast group. Messages sent by the client and target service are shown in Figure 9.11. Target services that match the probe send a response directly to the client. Clients can also find services by name. This can be done by sending a resolution request message to the same multicast group. Again, the target service that matches sends a direct response to the client. When a target service joins the network, it sends an announcement message to inform all clients of its presence. This minimizes the amount of unnecessary probing by the clients because they can now only listen to the multicast group to be notified of new services. When a target service leaves, it sends a bye message to the multicast group [9].

There is also alternative mode for the protocol that is aimed to minimize the multicast messaging. This mode requires presence of a new entity called discovery proxy. It can work as a messenger between clients and target services. It also sends an announcement message of itself when it detects a probe or a resolution request sent by multicast. However, the WS-Discovery protocol specification does not describe the discovery-specific protocol in detail and leaves it pretty much open.

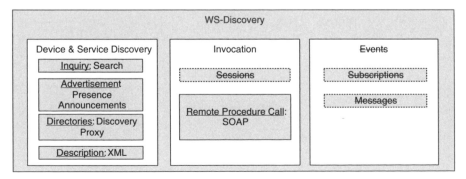

FIGURE 9.10   Web Services Dynamic Discovery toolkit template.

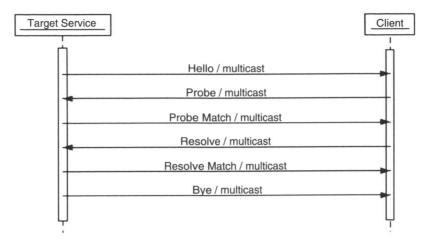

FIGURE 9.11 Message transfers between client and target service.

End points can implement more than one of client, target service, or discovery proxy entities [9].

All the messages in the WS-Discovery protocol are defined with XML descriptions in the Web services manner. Detailed descriptions of these messages can be found in the WS-Discovery specification. Messages are sent using SOAP over UDP [9].

9.4.9 eXtensible Service Discovery Framework

The eXtensible Service Discovery Framework (XSDF) [10, 38] is an evolution of the Service Location Protocol (SLP). It is designed to address several problems of SLP [39]. It is very new and currently in the state of Internet-Draft (I-D). A total of seven I-Ds were just released. XSDF introduces many improvements over the SLP protocol. Figure 9.12 shows the XSDF framework. Most important features of XSDF are

- *Enhanced service model*: Services are not identified by URL as in SLP but by a new advanced concept involving UUIDs.
- *Internet-wide location*: SLP was designed for local-area networks within a single administrative domain. XSDF is designed to be used for Internet-wide service discovery.
- *Load balancing and high-availability*: For the purpose of becoming Internet-wide solution for service discovery, XSDF also incorporates new features that can be used to select the best service and distribute data of registered services.

Similar to SLP architecture, XSDF architecture consists of User Agents (UA), Service Agents (SA), and Directory Agents (DA). XSDF does not change their fundamental properties, it just extends their functionality and tries to achieve architectural improvements.

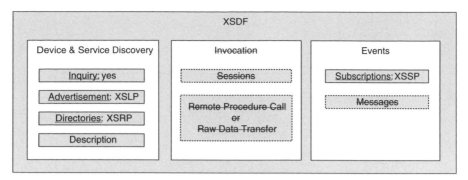

FIGURE 9.12 eXtensible Service Discovery framework.

These XSDF Agents communicate with four protocols [10]:

- *eXtensible Service Location Protocol (XSLP)*: XSLP is the protocol used in the actual service discovery. UAs use it to find services from DAs or directly from SAs [39].
- *eXtensible Service Register Protocol (XSRP)*: XSRP is used by the SAs to register and refresh their service information to DAs [40].
- *eXtensible Service Subscription Protocol (XSSP)*: XSSP is used by XSDF Agents to subscribe to get notifications of changes in service information; typically used between DAs [41].
- *eXtensible Service Transfer Protocol (XSTP)*: XSTP is used to achieve scalability and high-availability between DAs [42].

9.5   IMPROVING SERVICE DISCOVERY

We looked at adding directories to protocols that do not have them, for example UPnP. We looked at using location awareness to improve discovery speed, improve approximation of nearness, and enable remote access. We looked at implementing proxies in the fixed network to help cope with the intermittent connectivity of wireless devices.

9.5.1   Security

Security is an important issue in Service Discovery. The common situation is that it is assumed that all the devices are on a private network that is isolated from any threats by a firewall. This can be achieved in the home by using a wired network or by properly configuring the WLAN security settings.

Mobile computing introduces new complications because a mobile computer frequently encounters unfamiliar networks. How does the homeowner give a visitor access to the home network? How do you take it away? A detail-oriented network administrator can easily handle these tasks, but we believe it should be as easy as operating a television set.

Authentication issues become import in remote access scenarios for the extended home.

We don't have answers to present here. Many people are working on many solutions—and mechanisms are a topic of debate in standardization groups.

9.5.2 Semantics and Automatic Composition

There is not very much intelligence in so-called Smart Spaces. All the intelligence must come from the *user*. This is exactly the opposite of the desired situation. The research question is "Can we somehow automate the intelligence involved in combining devices into useful configurations?"

The devices in a smart space should be able to describe themselves sufficiently enough to automatically combine their capabilities into useful configurations.

There is research being done to add semantic information to device descriptions and service interfaces. The idea is that a computer program automatically finds a device that will serve a specific purpose. Semantic Web researchers are studying "service composition" by which Web Services are automatically combined. If we can automatically determine what a device can do and how it is used, then some sort of Artificial Intelligence can be used to compose services.

The interested reader should follow up be researching Semantic Web Services, OWL-S and "service composition."

9.5.3 Interoperability

Because the world is not going to be united under a single distributed middleware toolkit any time soon, there is a need for interoperability between protocols. Exactly how the interoperability is achieved is an open question. Should every device speak multiple protocols or are there protocol translators in the network?

9.5.4 Touch

One way to introduce a new device into a home network is with a touch interface. A touch interface allows the user to physically touch a device to another to initiate communication. This can be done with Radio Frequency Identification (RFID) tags, infrared messaging, or USB cables.

9.5.5 Directories

In the context of discovery protocols, directories are caches of services that have already been discovered. These services are remembered for the purpose of improving the speed and reliability of the discovery process. The discovery process can be quite time consuming—in UPnP it can take several minutes. The discovery process can also be unreliable; multiple queries may be needed before all devices are discovered. A mobile device that has gone into sleep mode may[1] not be discoverable at all. Directories can provide the benefit of immediately providing a list of local devices, bypassing the discovery process.

One disadvantage is that directories can give false positives if they fail to notice that a device has left. Another disadvantage is that if directories become a requirement, then discovery cannot operate in their absence. SLP has the best practice—directories are used, but on a network that lacks a directory, SLP devices will use direct peer-to-peer discovery.

9.5.6   Location Awareness

Location awareness can be used to improve the usability and efficiency of discovery in several ways. Usability can be improved by organizing discoveries according to their locations. Efficiency can be improved by assuming that devices have gone out of range when the user changes location. Another means to improve both efficiency and usability is to remember the discoveries in each location then using that information to instantly populate a user interface. Testing for the presence of known devices can be faster than device query mechanisms.

People naturally take spatial relationships, such as physical proximity, into account when dealing with or organizing objects in the physical world. Integration of location awareness with service discovery would make it possible for applications to reason about the location and proximity of networked devices and services. The ability of applications to use discovery to reason about the physical location of networked devices and services provides a powerful and intuitive tool for humans to manage and interact with "smart spaces" or other mobile computing environments.

Integration of location awareness with service discovery raises some interesting issues: How can location be determined? How should location information be represented? How should queries about location be expressed? What is the best way to trigger actions in response to a change of location? What sorts of protocols are needed to deal efficiently with frequent changes of location? How can privacy of a user's location be preserved?

All discovery protocols have a means to advertise a device and a means to discover available devices. Once a device is discovered, there is a means to retrieve a description of the device and its services. We have developed similar location aware prototypes of SLP and UPNP. Here we will focus on our work with UPNP.

9.5.7   Service Browsing

The predominant interactions model is to search for services when they are needed. The user starts with an application that requires some other service—the application uses discovery to find a satisfactory instance of the service.

---

[1]However, a feature called "wake on LAN" offers a potential solution. The problem is that too many messages can wake up a sleeping device enough times to drain the battery.

We explored a different model of user interaction named *Service Browsing*. In Service Browsing, the user starts by discovering all the services that are available. Next, the user chooses a service. The Service Browser launches the appropriate application—or displays a list of appropriate choices. If there is no application, then the Service Browser may try to download one.

9.5.8 Proxies

We created proxies that represent a device when it has become unreachable. Intermittent connectivity is an inescapable attribute of wireless networks. The causes of intermittent connectivity include (1) wireless devices frequently go into sleep mode to conserve battery power, (2) wireless devices can be carried out of range, (3) wireless connections are subject to interference from other sources.

The proxies programs live on the fixed network. They act on behalf of a missing wireless device in at least two ways: They represent its presence by responding to discovery inquiries. They store messages accumulated during the device's absence and replay them when the connection is restored.

9.6 CONCLUSIONS

Today's mobile computers include additional features such as Java, Bluetooth, Smart Covers, WAP 2.0, and JavaScript enabled Web pages and with embedded SOAP, 3GPP with SIP, and other technologies that make it possible to provide sophisticated, distributed applications. These new applications need Discovery tools to learn about the nearby networks or network-accessible resources available to them.

Mobile computers will be even more dependent on discovery technology than today's laptop and desktop systems. Desktop and even laptop computers are not very mobile. Even laptop computers are usually connected to a small number of familiar networks—usually the office network and the home network. Unlike laptop computers, mobile terminals equipped with Bluetooth or WLAN networks will be constantly connecting and disconnecting from networks.

Based on the above, discovery is an essential building block for self-configuring systems in home networks. In this chapter, we described that all discovery protocols have some facility for inquiry and response. Other common features are presence advertisements, directories, and self-description. Discovery protocols are normally part of a distributed middleware toolkit. These toolkits usually include facilities for remote invocation and events.

We looked at some of the most well established protocols and discussed them in the context of home networks. In home networks, you can assume that all devices will be within multicast range. Yet, a home network can also extend beyond the home in two different ways: (1) there may be an overlay network that makes a distributed set of devices appear to be on the same local network, and (2) some devices access services on the Internet such as television program guides.

There is enough information here to begin designing a new discovery protocol.

However, technology choices can be independent of market factors only when there is no need to integrate with existing devices and when there is no need to deploy into environments based on established discovery protocols. If you want to interoperate with devices from other manufacturers, then you must choose a discovery protocol that has successful market adoption such as UPnP or Bonjour. You can further increase your odds of market acceptance by supporting multiple protocols.

Improvements to discovery protocols are needed. Issues related to mobile computing are one of the most important and interesting areas for researchers to address. Another neglected area of discovery protocols is security. Researchers have many good opportunities for making improvements to discovery protocols.

REFERENCES

1. http://www.ietf.org/rfc/rfc2608.txt, "Service Location Protocol, Version 2," E. Guttman, C. Perkins, J. Veizades, M. Day, IETF RFC 2602, June 1999.

2. http://developer.apple.com/networking/bonjour/index.html, "A Hybrid Approach for Location-based Service Discovery in Vehicular Ad Hoc Networks," N. Klimin, W. Enkelmann, H. Karl, and A. Wolisz.

3. http://www.upnp.org/download/UPnPDA10_20000613.htm, "Universal Plug and Play Device Architecture",Version 1.0, 08 Jun 2000, Microsoft.

4. http://wwws.sun.com/software/jini/, "The Jini Specification," Ken Arnold, Bryan O'Sullivan, Robert W. Scheifler, Jim Waldo, Ann Wollrath.

5. JXTA: Java P2P Programming (ISBN: 0-672-32366-4), D. Brookshier, D. Govoni, N. Krishnan, and J. C. Soto.

6. http://www.jxta.org (Home page of Sun's JXTA project).

7. http://www.ietf.org/rfc/rfc2131.txt, "Dynamic Host Configuration Protocol," D. Roms, March 1997, IETF RFC 2131.

8. http://www.bluetooth.org/docs/Bluetooth_V11_Core_22Feb01.pdf, "Bluetooth V1.1 Core Specifications".

9. http://msdn.microsoft.com/ws/2005/04/ws-discovery/, "Web Services Dynamic Discovery."

10. M. Uruena and D. Larrabeiti, Overview of the eXtensible Service Discovery Framework (XSDF). Internet-draft, Internet Engineering Task Force, March 2004.

11. http://mit.edu/people/elliot/thesis.ps, "Design and Implementation of Intentional Names," Elliot Shwartz, Master's Thesis, MIT, May 1999.

12. http://ninja.cs.berkeley.edu/, (Home page of the UCB Ninja project).

13. http://www.freenetproject.org, (FreeNet project home page).

14. http://www.cooltown.hp.com/cooltownhome/, (This is the home page of the HP Labs CoolTown research project).

15. http://www.salutation.org, (Home page of the Salutation Consortium).

16. http://www.syncml.org, (Home page of SyncML, Inc.).

17. ITU-T Rec. X.500, "The Directory: Overview of Concepts, Models and Service," 1993.

18. http://www.ietf.org/rfc/rfc2251.txt, "Lightweight Directory Access Protocol (v3)," M. Wahl, T. Howes, and S. Kille, IETF RFC 2251, December 1997.

19. http://www.uddi.org/specification.html, (The UDDI web page pointing to version 1 & 2 of the UDDI specifications and a collection of supporting documents).

20. C. Andrew, B. Brown and B. Bodine, Novell's NDS Developer's Guide with CDRom, Hungry Minds, Incorporated, December 1998.

21. The Professional Reference, E. G. Laubach, B. Vines, New Riders Publishing, January 1994.

22. R. Orfali, D. Harkey and J. Edwards. Introduction to and Comparison of Corba and DCOM: The Essential Distributed Objects Survival Guide, John Wiley and Sons, Inc., New York, 1996.

23. W. Rosenberry, Understanding DCE, O'Reilly & Associates, September 1992.

24. http://www.ietf.org/rfc/rfc2543.txt, "SIP: Session Initiation Protocol," M. Handley, H. Schulzrinne, E. Schooler and J. Rosenberg, March 1999, IETF RFC 2543.

25. http://www.bluetooth.org/foundry/adopters/document/PAN_Spec_v1_0/, "Personal Area Networking Profile".

26. http://www.ietf.org/rfc/rfc1035.txt, "Domain Names—Implementation and Specification," P. Mockapetris, November 1987, IETF RFC 1035.

27. http://files.multicastdns.org/draft-cheshire-dnsext-multicastdns.txt, "Multicast DNS," Stuart Cheshire, Marc Krochmal, Apple Computer, Inc., June 2005.

28. http://www.ietf.org/rfc/rfc2782.txt, "A DNS RR for Specifying the Location of Services (DNS SRV)," A. Gulbrandsen, P. Vixie, IETF RFC 2782, February 2000.

29. http://www.ietf.org/internet-drafts/draft-ietf-dnsext-mdns-45.txt, "Linklocal Multicast Name Resolution," Bernard Aboba, Dave Thaler, Levon Esibov, Microsoft Corporation, October 2005.

30. http://www.zeroconf.org/Rendezvous/txtrecords.html, "DNS-SD (Rendezvous) TXT Record Format."

31. R. G. Golden III, Service Advertisement And Discovery, IEEE Internet Computing, 2000.

32. Sun Microsystems. Jini Architecture Specification 2.0, June 2003. Available at http://www.sun.com/software/jini/specs/jini2_0.pdf.

33. http://www.iana.org/assignments/bootp-dhcp-parameters.

34. http://www.ietf.org/rfc/rfc1001.txt, "Protocol Standard for NetBIOS Service on a TCP/UDP Transport: Concepts and Methods," A. Aggrawal et al., IETF RFC 1001, March 1987.

35. P. Bhagwat, Bluetooth, Technology For Short-range Wireless Apps. *IEEE Internet Computing*, 2001.

36. S. Helal, Standards For Service Discovery And Delivery. *IEEE Pervasive Computing*, 2002.

37. "Bluetooth Core Specification." Available at http://bluetooth.org/foundry/adopters/document/Bluetooth_Core_Specification_v1.2.

38. M. Uruena and D. Larrabeiti, extensible service discovery framework (xsdf): Common elements and procedures. Internet-draft, Internet Engineering Task Force, March 2004.

39. C. E. Perkins, "Service Location Protocol for Mobile Users." In *The Ninth IEEE International Symposium on Personal, Indoor and Mobile Radio Communications*, Volume 1, pp. 141–146, IEEE, September 1998.

40. M. Uruena and D. Larrabeiti, extensible service registration protocol (xsrp). Internet-draft, Internet Engineering Task Force, March 2004.

41. M. Uruena and D. Larrabeiti, extensible service subscription protocol (xssp). Internet-draft, Internet Engineering Task Force, March 2004.

42. M. Uruena and D. Larrabeiti, extensible service transfer protocol (xstp). Internetdraft, Internet Engineering Task Force, March 2004.

10

SMALL, CHEAP DEVICES FOR WIRELESS SENSOR NETWORKS

ZACH SHELBY, JOHN FARSEROTU, AND JOHN F.M. GERRITS

This chapter gives an overview of the technologies and issues facing the embedding of wireless sensor devices everywhere, especially applied to ubiquitous home networking. Wireless measurement and control is a very exciting aspect of home networking, with a lot of activity in academia and industry (as an example, the ZigBee effort). The core technical issues that are key in enabling such a vision involve the wireless interface itself, new system-on-a-chip technologies for truly cheap, integrated solutions, and finally on embedded software for these systems. The final goal of these combined is to enable creation of cheap, reliable, low-power, ubiquitous devices.

Current wireless embedded systems are implemented using multichip solutions, that is, they consist of multiple microcontrollers, Input Output (IO) processors, and a separate radio chipset. In addition, a large number of external passive and active components are required. Although the energy-to-performance ratio of individual chips is improving rapidly, the cost and total efficiency are still problematic. The most promising trend for the mass production of smaller, cheaper, lower-power devices is called system-on-a-chip. These mixed-signal Very Large Scale Integration (VLSI) chips integrate microcontrollers, IO capabilities, complete wireless transceivers, and even sensors. System designers can leverage global optimization, new manufacturing processes, and Micro Electro Mechanical Systems (MEMS) to optimize such chips.

Improvements in cheap, low-power, low-data-rate wireless techniques along with their global standardization is one of the keys to the large-scale deployment of wireless sensor networks. This is especially true for consumer devices related to body monitoring, personal communications, or home automation. Table 10.1 shows a

Technologies for Home Networking. Edited by Sudhir Dixit and Ramjee Prasad
Copyright © 2008 John Wiley & Sons, Inc.

TABLE 10.1 Comparison of Low-Data-Rate Wireless Solutions

Parameter	IR-UWB 802.15.4a (4 GHz)	FM-UWB (4 GHz)	433/867 MHz (XE1203)	Bluetooth (2.4 GHz)	802.15.4 (2.4 GHz)
Power consumption	5–20 mW*	Tx = 3.5 mW†, Rx = 7.5 mW	Tx = 80 mW, Rx = 34 mW	100 mW	Tx = 25–50 mW, Rx = 60 mW
Radiated power	−41.3 dBm/Hz	100 μW	3 mW	1 mW (Power class 3), 2.5 mW (Power class 2), 100 mW (Power class 1)	100 mW (Europe) 1 W (USA)
Data rate	100 kbps to 10 Mbps	1–100 kbps	Up to 62.5 kbps	723.2 kbps	250 kbps
Multiple users	Up to 255 devices per piconet	15 @ 100 kbps, 150 @ 1 kbps	2 in FDMA mode, others in TDMA mode	8 active users	Up to 255 devices per piconet
Medium access	CSMA-CA with TDMA-based beaconing	FDMA subcarrier and TDMA	FDMA or TDMA	FHMA TDMA/FH combination	CSMA-CA with TDMA-based beaconing
Robustness	Highly multipath and interference resistant band change	Analog spread spectrum, multipath diversity	Frequency or band change	Frequency hopping, block code	Symbol-to-chip mapping (32-chip PN sequence)
Range	1–10 m	1–10 m	10–100 m	10 m, 20 m, 100 m	10–100 m
Additional features	ToA estimation for accurate positioning, coexistence	RSSI, scalable RF BW, coexistence, spectrum, rapid synchronization	RSSI	RSSI	RSSI
Status	Prototype	Prototype	ISM band module	Commercial-off-the-shelf	Commercial-off-the-shelf

*Estimation based on prototype chips from the EU Pulsers project.
†Continuous power consumption at 100 kbps. Lower power possible dependent on the duty cycle.

comparison between low-data-rate wireless technologies. Traditionally, proprietary Industrial Scientific Medical (ISM) band wireless modems have been used for embedded telemetry and automation solutions. For personal communications, Bluetooth [1] is used heavily, for example with mobile phones and PDAs. It is, however, too power hungry and topology limited for sensor networking and automation. In 2005, the IEEE 802.15.4 standard [1] for low-power WPAN communications was released and has been rapidly commercialized. This wireless interface standard is sometimes referred to as ZigBee [2], which is one possible protocol stack solution over the standard. IEEE 802.15.4 provides the first global standard for wireless sensing and control and offers huge potential.

Although the IEEE 802.15.4 standard has been a breakthrough for sensor networking, there is still room for improvement. For example, there is still a need for even cheaper, smaller, and lower-power wireless technology, with added-value features such as robustness to interference and multipath and accurate ranging capabilities. Ultra-wideband techniques are providing the most recent answer to these demands. Although widely known in North America for superhigh-data-rate wireless USB solutions, UWB can also be applied to superlow-power, low-data-rate solutions. The Swiss Center for Electronics and Microtechnology (CSEM) has developed a solution called frequency modulation UWB (FM-UWB), which will be introduced in more detail in Section 10.3. A more traditional approach is to use very narrow pulses, which result in very wide bandwidth. This technique is called impulse response UWB (IR-UWB). An experimental implementation of IR-UWB is introduced in Section 10.1. More recently, this IR-UWB technique has been adopted for the first extension to the IEEE 802.15.4 standard, which will be examined in the next section.

10.1 IMPULSE RADIO UWB

Ultra-wideband is a promising technology for wireless sensor networking for many reasons. First, it is very robust to multipath effects and to interference, useful in for example indoor home networking environments. Very simple transceiver architectures can be designed, enabling cheap, low-power implementations, and at the same time, higher data rates can be easily achieved compared with typical narrowband or spread-spectrum technologies. Finally, impulse-response UWB uses extremely narrow pulses to achieve such a large bandwidth, thus giving very accurate time-of-arrival capabilities for positioning. In general, the accepted definition for UWB technology is given by the Federal Communications Commission (FCC), stating that it is any scheme that occupies a fractional bandwidth $W/f_c = 20\%$, where W is the transmission bandwidth and f_c is the band center frequency, or more than 500 MHz of absolute bandwidth [3].

Impulse radio–based UWB (IR-UWB) systems have great potential in low-data-rate applications such as sensor networking and automation for the home. The advantages from UWB arise from the nature of the signal transmission. UWB transmitters produce very short time-domain pulses that can propagate without an additional RF mixing stage. These narrow pulses produce a very wide band, which is a noise-like signal and is resistant to severe multipath channels and jamming. In addition, very good time-domain resolution allows for location and tracking capabilities [1]. Applied to wireless sensor networking UWB allows for energy savings through lower duty cycles as high data rates can be achieved with relatively low power consumption when compared with narrowband solutions. In this section, we look at experimental work done on a low-data-rate UWB solution for wireless sensor networks.

The Centre for Wireless Communications has developed an efficient, low-cost, low-complexity concept for UWB communications. The application space for the design is for low-data-rate applications such as sensor and embedded networking where location and tracking capabilities are valuable. The experimental design has been implemented in an application-specific integrated circuit with a 0.35μ BiCMOS Si-Ge process (see Fig. 10.1). A noncoherent scheme is used with bit position modulation (BPM) and direct sequence (see [4] for a detailed paper on the design). A pulse generator sends pulses of approximately 350 picoseconds in duration repeated at a high rate. A noncoherent energy collection technique is used for the receiver including time-of-arrival (ToA) estimation.

A system using noniterative positioning has been developed for use with the UWB transceivers [5]. Better ToA accuracy could be achieved with shorter integration times, a high sampling rate, or a separate receiver branch for ToA estimation, although at the cost of added complexity. Figure 10.2 shows a block diagram of the transceiver. The experimental ASIC currently implements the analogue parts of the receiver, with the baseband processing contained in an FPGA. A 5-Mbps data rate is the default for the design. This data rate, although high for a narrow-band transceiver, is very efficient with an IR-UWB design. In addition, it allows low-power

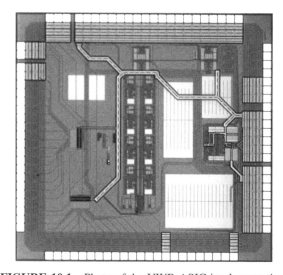

FIGURE 10.1   Photo of the UWB ASIC implementation.

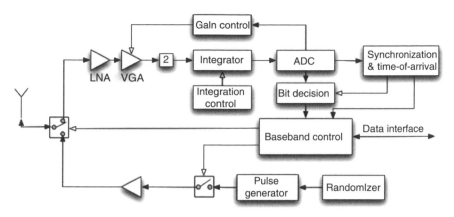

FIGURE 10.2 Transceiver block diagram.

nodes to achieve very low duty cycles, thus reducing power consumption. Considering a device data-rate target of several kbps, an aggregate of 5 Mbps means many hundreds of devices may be operational in the system.

Time-division multiple-access (TDMA) and power-sense multiple-access (PSMA) schemes can be used with this noncoherent UWB solution. TDMA has advantages for systems needing highly accurate positioning and real-time communications, however the ad hoc flexibility of PSMA has its own advantages. Hybrid solutions are also possible, with contention periods within TDMA. As seen in Section 10.1, IR-UWB in the context of IEEE 802.15.4a standardization uses such a hybrid approach. The current MAC implementation is a TDMA solution using a clustered access point (AP) approach for easy mobility. See Figure 10.3 for the TDMA frame structure. To support the low-power, low-cost requirements, the MAC solution must be low complexity and support an environment with a minimum of inter-user

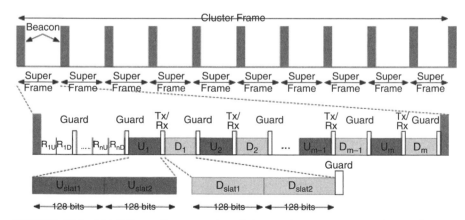

FIGURE 10.3 MAC Frame Structure where R is a random access slot, U is an uplink slot, D is a downlink slot, Guard is the guard time, and Tx/Rx is the turnaround time.

interference, which is important when attempting to undertake positioning measurements. For similar reasons, the system being developed employs time division duplexing (TDD). There is an "up-link" time period when the low-cost UWB sensors can send information to fixed nodes in the network, and there is a "down-link" time period where the UWB devices can receive commands and information. Each time, the frame is introduced by a beacon that carries information on the availability and the structure of the network.

To aid the mobility of devices while maintaining the low-complexity requirement of the sensors, the system is configured to support handover of UWB sensors as they move through the network. Cluster-frames are split into 10 superframes, each of which is rotated between access points in a cluster. In each superframe, a node is assigned a number of uplink slots, immediately followed by downlink slots. Once a node enters an area of UWB APs, it is assigned slots for communication and can roam throughout an area of APs with minimal need for control communications. The system is designed to keep nodes as simple as possible. Peer-to-peer and multi-hop communications are achieved with contention periods. However, our UWB solution can also be used with other MAC techniques. One of the biggest potentials for this type of low-cost UWB technology is in inclusion to the IEEE 802.15.4a standard. This will provide a physical interface option with higher data rates, lower power consumption, and positioning capabilities compared with the current 868 MHz and 2.4 GHz options. More about the IEEE 802.15.4a standardization process can be found from Section 10.2.

10.2   IEEE 802.15.4A

To satisfy the demands for extended range and location capabilities of the low-rate 802.15.4 standard, the IEEE recently established the 802.15.4a task group [6] to explore an alternative physical (Alt PHY) layer concept for low-data-rate applications. The working group, IEEE 802.15 TG4, is chartered to investigate low-data-rate solutions for very-low-power and very-low-complexity systems. It is intended to operate in unlicensed, international frequency bands. Potential applications include sensors, interactive toys, smart badges, remote controls, and home automation, as well as safety and industrial applications from fire detection to smart warehousing and building automation. Although not specifically targeted at UWB, the promising physical layer properties of the technology ensured that the candidates under discussion were almost exclusively based on IR-UWB. The working group has decided upon supporting both IR-UWB and a 2.4 GHz chirp spread-spectrum technique; here we will describe the main features of the IR-UWB proposal.

The draft specification currently defines approximately 500 MHz UWB channels from 3.1 to 4.9 GHz and an upper band from 4.9 to 10 GHz. Thus by default, three channels will be available, although only one channel is mandatory. A default data rate of 842 kbps has been defined, along with optional rates between 100 kbps and 27 Mbps depending on the channel. The draft also includes standard mechanisms for performing accurate ranging. The new wireless interfaces make use of the same

IEEE 802.15.4 MAC, shared in common with the base 802.15.4 standard. Support for noncoherent UWB receivers is also included in the draft, making a solution quite close to that described in Section 10.1 viable for implementing the upcoming standard.

10.3 FREQUENCY MODULATION UWB

Frequency Modulation Ultra-Wideband (FM-UWB) is a low-complexity UWB solution that offers ultralow power, yet robust performance for short-range, low-data-rate (LDR) wireless communication. Many wireless communication systems are capable of supporting operation at low data rates. However, this does not make them low complexity and low power.

All UWB systems are subject to strict regulatory limitations on their radiated power. Hence, the power consumption of the radio transmitter tends to be relatively low. The same is not true for the receiver, where substantial processing power may be required for implementation of, for example, IR-UWB based solutions (e.g., synchronization). The simplicity of the FM-UWB approach sets it apart. The low-complexity implementation illustrated by Figure 10.2 helps enable FM-UWB to be not only low radiated power but also low power consumption.

Today, the FM-UWB solution is a prototype. IC building blocks are being developed in the European IST MAGNET Beyond Integrated Project [7]. From Table 10.1, it can be seem that the radiated power is of the order $100\,\mu W$. This is based on -41.3 dBm/MHz and a bandwidth of 500 MHz. The estimated power consumption is 3.5 mW for the transmitter and 7.5 mW for the receiver (i.e., continuous power, 4 GHz band implementation). This compares favorably with commercial narrowband solutions that lack the inherent robustness of FM-UWB to the interference and multipath. The data rate is $1-100$ kbit/s, which is suited to health and activity monitoring applications, which are typically not more than a few kbit/s.

Figure 10.4 provides a block diagram of the FM-UWB transmitter [8]. Digital data $d(t)$ is modulated on a low-frequency subcarrier (typically $1-2$ MHz for 100 kbit/s data) using Frequency Shift Key (FSK) modulation with a modulation index of $\beta_{SUB} = 1$. The modulated subcarrier signal m(t) then modulates the RF oscillator, for example, at $f_C = 4$ GHz with $\Delta f = 800$ MHz yielding the constant envelope UWB signal.

Figure 10.5 is an illustration of the data, the subcarrier, and the UWB signal in the time domain for a data transition at $t = 0$ and sub-carrier frequency of 1 MHz. For

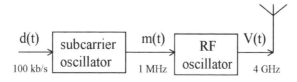

FIGURE 10.4 Block diagram of the FM-UWB transmitter.

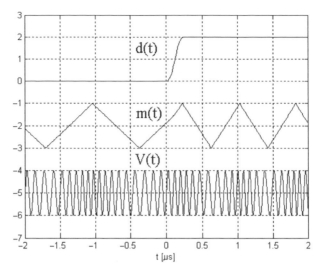

FIGURE 10.5 Time domain view of data $d(t)$, subcarrier $m(t)$, and UWB signal $V(t)$.

ease of illustration, the center frequency of the UWB signal $V(t)$ was chosen to be 10 MHz.

The choice of the subcarrier frequencies f_{SUBi} and modulation indices β_{SUBi} is determined by the data rate(s) and the number of users in the UWBFM communications system. A low subcarrier modulation index yields a lower subcarrier bandwidth allowing more users.

An example of a hybrid system providing multiple low-data-rate (LDR) channels is shown in Figure 10.6. In this example, three LDR users (1–3) each operate at

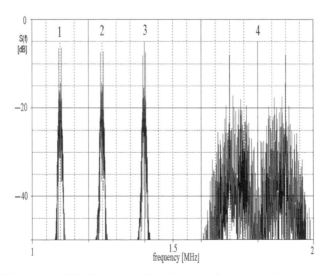

FIGURE 10.6 A hybrid LDR system with subcarrier frequencies between 1 and 2 MHz.

10 kbit/s. Given a modulation index $\beta_{SUB}=1$, the bandwidth is 20 kHz. The users are spaced 150 kHz apart. The fourth user (4) operates at 100 kbit/s and uses a modulation index $\beta_{SUB} = 2$ and a bandwidth of 300 kHz.

When Direct Digital Synthesis (DDS) techniques are used for the subcarrier generation, a low-complexity yet high-performance and flexible transmitter can be realized. The subcarrier signal $m(t)$ modulates the RF Voltage Controlled Oscillator (VCO) operating in open-loop mode at the center frequency (f_c) of the UWB signal. The VCO tuning curve is periodically measured and stored by a frequency synthesizer–based calibration system. This allows minimization of the transmitter power consumption. Moreover, due to the high deviation of the FM-UWB signal, phase noise specifications for the transmitter VCO are extremely relaxed, allowing the use of simple and low-power hardware. As such, the hardware implementation for FM-UWB is potentially very low power and low cost compared with other wireless schemes due to the relaxed oscillator phase noise requirements for the transmitter and, as we shall see, the absence of any frequency conversion in the receiver.

The FM-UWB receiver demodulates the RF signal without frequency translation. No local oscillator and no carrier synchronization are required. Figure 10.7 shows the block diagram of the FM-UWB receiver. It comprises a Low Noise Amplifier (LNA), wideband FM demodulator, subcarrier filtering, and one or several amplification stages and subcarrier demodulators.

The wideband FM demodulator is implemented as a delay line demodulator as shown in Figure 10.8. Both center frequency and bandwidth of this demodulator are tunable and tailored to the received FM-UWB signal. Codesign of LNA and wideband FM demodulator further helps to lower receiver front-end power consumption.

The power spectral density (PSD) of a wideband FM signal is determined by and has the shape of the probability density function of the modulating signal $m(t)$. Triangular subcarrier waveforms have a uniform probability density function and therefore yield a flat RF spectrum. Figure 10.9 provides an example of the power spectral density of a FM-UWB signal obtained with a triangular subcarrier: a flat RF spectrum with steep spectral roll-off that—without any additional filtering—fits the FCC/ETSI spectral mask.

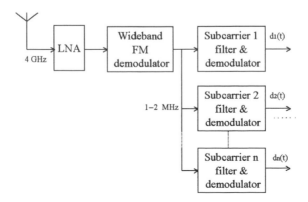

FIGURE 10.7 Block diagram of the FM-UWB receiver.

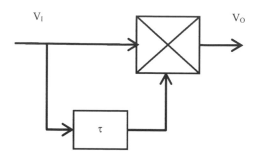

FIGURE 10.8   Wideband FM demodulator.

Instead of filling the complete lower band from 3.1 to 5 GHz with a single FM-UWB signal, Figure 10.10 shows that is also possible to generate a FM-UWB of lower bandwidth (e.g., 500 MHz) to avoid interference to and from, for example, WiMAX systems. If the appropriate technology is used for IC implementation (e.g., Si-Ge), the FM-UWB system can also be implemented between 7 and 10 GHz.

FM-UWB may be seen as an analogue implementation of a spread-spectrum system with spreading gain equal to the ratio of RF and sub-carrier bandwidth

$$G_{\text{PdB}} = 10 \log_{10} \left(\frac{B_{\text{RF}}}{B_{\text{SUB}}} \right) = 10 \log_{10} \left(\frac{2\Delta f_{\text{RF}}}{(\beta_{\text{SUB}} + 1)R} \right)$$

In a 100 kbit/s LDR system with a RF bandwidth of 1 GHz, a processing gain of 37 dB is obtained. As an ultra-wideband signal, FM-UWB offers diversity in

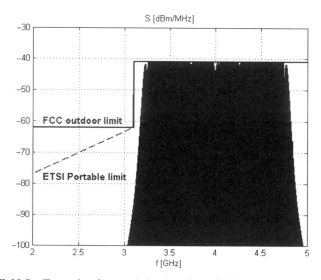

FIGURE 10.9   Example of spectral density of a 1.6-GHz-wide FM-UWB signal.

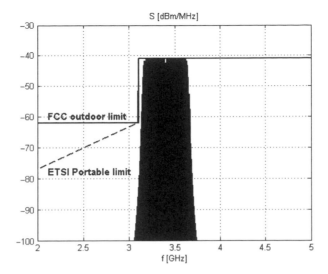

FIGURE 10.10   Example of spectral density of a 500-MHz-wide FM-UWB signal.

presence of frequency selective multipath fading. Importantly in the case of FM-UWB, this is accomplished without the need for sophisticated baseband processing (e.g., OFDM, adaptive equalization), as subcarrier detection is effectively narrowband (typically 200 kHz in a 100 kbps FM-UWB system).

FM-UWB may be used with either standard or proprietary Medium Access Control (MAC) solutions. An IEEE 802.15.4 compatible MAC is being implemented for use with FM-UWB in MAGNET Beyond for BAN/PAN applications. This standard MAC offers a common interface to upper layers. It is also possible to use the FM-UWB air interface in a Frequency Division Multiple Access (FDMA) like mode. In this mode, the MAC is very simple and processing is limited. As such, ultra-low-power performance, potentially compatible with energy scavenging (e.g., less than approx. 10–100 μW average power) is foreseen for BAN/PAN applications employing star network architectures (refer to Chapter 11). Additionally, adaptation to use with CSEM's ultralow-power WiseMAC for wireless sensor network applications is being investigated [9].

Although a proprietary solution today, efforts are under way to standardize FM-UWB. Both ETSI and the IEEE are targeted (e.g., ETSI TG31a and the IEEE 802.15.4 Interest Group BAN). The FM-UWB standardization activities are supported by the MAGNET Beyond Standardisation Forum.

10.4   SYSTEM-ON-A-CHIP

The success of wireless sensing and control technology depends heavily on the ability to highly integrate these wireless embedded systems on cheap silicon, produced in large quantities. The trend toward system-on-a-chip (SoC) implementations is currently very promising. System-on-a-chip refers to the very-large-scale integration (VLSI) application-specific integrated circuit (ASIC) implementation of an entire system. In the case of wireless sensor networks, we are interested in the integration of a complete radio transceiver, a microcontroller, IO interfaces, and possibly even sensors. New silicon fabrication processes, micro-electro-mechanical systems (MEMS), mixed-signal designs, and design processes are key enablers in the move toward system-on-a-chip.

Early system-on-a-chip solutions for wireless sensor networks have begun to hit the market, with the integration of IEEE 802.15.4 radio interfaces with microcontrollers powerful enough to implement a sensor node implementation. One example of this is the Texas Instruments CC2430 chip, which integrates a 2.4 GHz IEEE 802.15.4 radio interface including hardware protocol support, and a fast 8051 microcontroller [10]. A minimal amount of external components are needed to create a stand-alone node—just a PCB antenna, a clock crystal, some passive components, and power. Digital and analogue IO is provided on the chip, so sensors can often be directly connected. The research community is also working very hard on this subject and are working toward even smaller, cheaper, and lower power SoC implementations. One good example is the Picoradio project from the UC Berkeley [10]. At first release, the CC2430 chip, and other similar ones, are available for under 5 USD in low quantities (1k), with the price decreasing quickly. The chip is about 5×5 mm in size. The Picoradio project advertises the goal of SoC technology for under 0.50 USD in quantity with an even smaller footprint.

10.5 EMBEDDED OPERATING SYSTEM

In addition to hardware, flexible and innovative software solutions play an important part in wireless sensor networks. As the computer industry has flourished from standardized, open operating systems supporting a wide variety of hardware platforms such as Linux, the same benefits can be achieved for wireless embedded systems for sensor networking. In the past, most tiny embedded systems have been implemented in C or assembler without an operating system, creating new software for each hardware platform and application. Open source, portable embedded operating systems and protocol stacks specifically aimed at wireless sensor networking are now however becoming available. Such an operating system can support a large number of microcontroller architectures and hardware platforms while providing a uniform interface and communication protocols to the application developer. This shortens the time-to-market for developing wireless sensor system appreciably.

TinyOS [11] is a very compact event-based operating system designed for wireless embedded sensing applications. It has originated from the academic community in the United States and is distributed under the GPL license. The operating system uses a research language called NesC, which is compatible with GCC compilers. Flexible internal messaging systems, lots of example applications and tools, combined with support for many hardware platforms for sensor networking make it popular especially with universities. There have been a very large number of network protocols and radio chip drivers implemented for TinyOS as part of research projects around the world. The downside of TinyOS is its event-based nature, making it problematic for real-time applications.

The Free Real-Time Operating System (FreeRTOS) [12] is an open-source effort from the industrial control community and is distributed under GPL. This extremely small microkernel operating system is implemented in Ansi C, supporting a huge number of microcontrollers and compilers. This is a fully pre-emptive real-time operating system with independent tasks, memory management, and semaphores. Examples including an embedded TCP/IP stack for many demo modules have been developed, and the OS is in use in a wide range of embedded products. It is, however, only a kernel and does not provide networking or communications protocols. Wireless sensor networking support for FreeRTOS can be found from the NanoStack project from Sensinode [13], available open-source as GPL. NanoStack provides a reconfigurable architecture for the protocols needed in wireless embedded networking while providing a simple API to the application. The stack includes a variety of protocols such as nanoIP, 6LOWPAN, and IEEE 802.15.4 support.

10.6 CONCLUSIONS

Low-complexity wireless devices are essential for realization of low-cost, low-power solutions for use in the home networks of tomorrow. Such solutions may be employed by users in their home, or their home away from home, while on travel. Potential applications include extension of health and activity monitoring from wireless sensor devices in BANs, through home and mobile gateways, to health care professionals and service centers. For such applications, small, wearable, and potentially throwaway wireless devices are essential.

Today, health monitoring solutions remain relatively primitive. Wireless health sensor devices are neither very small, nor are they truly low power, low cost. Smaller, lower-cost, lower-power solutions are needed. The availability of ultralow-cost, low-power devices is considered to be a service enabler.

Communication is not the only application domain of interest with respect to home networks; localization and sensing are expected to play an increasingly important role in future ambient intelligent home environments. The availability of low-cost, low-power wireless devices requires an optimized solution taking into consideration the requirements from radio and protocol, to the operating system, application, and integrated circuit. Such devices are expected to play an important role in the realization of future ubiquitous home networks and ambient intelligent environments.

REFERENCES

1. IEEE Std 802.15.4^{TM}-2003: Wireless Medium Access Control (MAC) and Physical Layer (PHY) Specifications for Low-Rate Wireless Personal Area Networks (LR-WPANs), October 2003.
2. ZigBee^{TM} Alliance homepage. Available at http://www.zigbee.org.

3. S. Roy, J.R. Foerster, V.S. Somayazulu and D.G. Leeper, Ultrawideband radio design: the promise of high-speed, short-range wireless connectivity. *Proc. IEEE*, Vol. 92, pp. 295–311, Feb. 2004.

4. L. Stoica, A. Rabbachin, S. Tiuraniemi, H. Repo, and I. Oppermann, An ultra wideband system architecture for wireless sensor networks. *IEEE Trans. Vehicular Technol.*, Vol. 54, no. 5, pp. 1632–1645, Sept. 2005.

5. A. Rabbachin, L. Stoica, S. Tiuraniemi, and I. Oppermann, A low cost, low power UWB based sensor network. In *Proc. of IWUWBS/UWB-ST*, May 2004.

6. The IEEE 802.15 TG4a working group. Available at http://www.ieee802.org/15/pub/TG4a.html.

7. IST MAGNET Beyond Home Page. Available at http://www.ist-magnet.org.

8. J.F.M. Gerrits, M.H.L. Kouwenhoven, P.R. van der Meer, J.R. Farserotu, and J.R. Long, Principles and limitations of FM-UWB communications systems. *EURASIP Journal of Applied Signal Processing*, Vol. 2005, no. 3, pp. 382–396, 2005.

9. A-El- Hoiydi and J.D. Decotignie, Low power downlink MAC protocols for infrastructure wireless sensor networks. *Mobile Networks and Applications Journal*, Vol. 10, pp. 675–690, 2005.

10. J. Rabaey et al., PicoRadio supports ad hoc ultra-low power wireless networking. *IEEE Computer*, Vol. 33, no. 7, pp. 42–48, July 2000.

11. TinyOS Home Page. Available at http://www.tinyos.net.

12. FreeRTOS Home Page. Available at http://www.freertos.org.

13. NanoStack GPL Home Page. Available at http://www.sensinode.com.

11

"SPOTTING": A NOVEL APPLICATION OF WIRELESS SENSOR NETWORKS IN THE HOME

HENRY TIRRI

Wireless sensor networks are emerging as a revolutionary and critical information technology, and they are continuing the trend originating in mainframe computing currently at the stage of mobile computing. This trend shows several aspects consistent in the evolution of computing including the increasing miniaturization of the computing units and an increasing emphasis of the role of "plug-n-play" communication between the computing units—"networking." In addition, from the software side there is an increasing need to software solutions that are robust, exhibit distributed control, and provide for collaborative interfaces, resulting in adaptive capabilities at the system level.

Like the current Internet, sensor networks are large-scale distributed systems but composed of smart sensors and actuators. Sensor networks will eventually infuse the physical world and provide "grounding" for the future Internet as one can use the semiconductor manufacturing techniques that underlie this miniaturization to build radios and exceptionally small mechanical structures that sense fields and forces in the physical world. These inexpensive, low-power communication devices can then be deployed throughout a physical space, providing dense sensing close to physical phenomena, processing and communicating this information, and coordinating actions with other nodes. Combining these capabilities with the system software technology that forms the current Internet makes it possible to instrument the world with increasing fidelity and will result in a "ubiquitously present" next-generation Internet. Sensor networks will form a critical infrastructure

Technologies for Home Networking. Edited by Sudhir Dixit and Ramjee Prasad
Copyright © 2008 John Wiley & Sons, Inc.

resource for society as they will monitor and collect information on diverse subjects as ecosystem dynamics, soil and air contaminants, medical patients, buildings, bridges, and other man-made structures. In this chapter, we limit our discussion to the role of wireless sensor networks in the home environment.

Research on wireless sensor networks has been taking place at several levels, from the lowest physical level to the highest information level; however, the latter is much less developed than the research at the physical levels. This has sometimes led to the misconception that the sensor network research is only about the hardware-level design (sensor design, physical and functional characteristics, semiconductor manufacturing techniques) and offers only additional input to traditional systems (see, e.g., the discussion of "context awareness" in mobile computing research). Albeit somewhat true at the current stage of research and experimentation, this is a very limited view of the fundamental changes in computing, computing platforms, and, at higher levels, about the "intelligence" of computing facilities.

Much of the research in wireless sensor networks has been focusing on military or environmental applications. However, they can also play an important role in the realization of ubiquitous computing for everyday life. Home environment is particularly suitable for the deployment of wireless sensors as some of the typical challenges, such as energy harvesting and security, are not so severe in the home domain as, for example, in the environmental applications. In addition, at homes there are many natural gateways to collect and process the sensor information— static ones such as media devices or a personal computer, or mobile devices such as smart phones that can collect sensor information when entering the communication range of an active sensor. Furthermore, use of mobile devices allows natural means for remote monitoring of home while, for example, traveling. The natural applications of wireless sensor network technology at home include, in addition to the surveillance functions discussed above, adding "intelligence" to utility consumption, electronic tagging, contamination control, disaster monitoring, status monitoring, and remote control.

One area of application of sensors and sensor networks that has received unanimous attention is that of In-Home Health Care, driven primarily by the ever-rising cost of health care and the patients wanting more control of their own monitoring and lifestyle. Just based on the basic demographic data, over the next decade, the number of people reaching age 65 will grow dramatically, and the costs of caring for age-related conditions will likely increase.

Intel's prototype sensor networks "Sensing Social Health" demo looks for sudden declines in social contact, tracks visually a person's daily interaction with others through sensors embedded throughout the home, and employs a screen phone that uses the sensor-delivered data to provide rich contextual cues, such as who is calling, when the parties last spoke, and what was discussed. Another demo, called "Caregiver's Assistant," demonstrates a smart home system that detects, monitors, and records the daily living activities of an elder by collecting data through postage stamp–sized wireless Radio Frequency Identification (RFID) tags affixed to household objects. Ultimately, the system could help manage everyday activities so that the elders' independence is maintained while relieving some of the burden of around-the-clock care by caregivers. In the sequel, we will not discuss this interesting area further but focus on the use of wireless sensor networks in "everyday life" at home. This restriction was deliberate as we feel that good motivational concept papers for that domain are already available. Therefore, our discussion in this chapter will be more from the viewpoint of context-aware computing at home as enabled by the deployment of wireless sensor networks. This is an area that is just beginning to be explored.

11.1 HETEROGENEOUS WIRELESS SENSOR NETWORK ARCHITECTURE

Much of the wireless sensor networks research has concentrated on rather homogeneous networks (i.e., networks where the nodes are very similar in their resources). This is natural if one considers the military and environmental application drivers but is not at all a good fit for consumer-oriented applications at home. In the home environment, wireless sensor networks will typically consist of a set of heterogeneous nodes with dynamic bindings—networks exist only when and where needed. Naturally, some of the subnetworks, such as the embedded sensors for monitoring utility consumption, temperatures, and motion detection, are very static in nature, but even such networks are presenting interfaces for dynamic nodes for control purposes. Therefore, the reality for deployed wireless sensor networks will be analogous to the developments in data management systems where the homogenous, single data model approaches have been replaced by so-called "federated" architectures where partially autonomous nodes with heterogeneous data models are cooperating with common protocols (Internet Protocols and World Wide Web being the most prominent examples).

For this type of a heterogeneous environment, an embedded, autonomic peer-to-peer model based on localized and occasional information exchange, and supporting a large range of device capabilities, ranging from very cheap tiny devices such as sensors, actuators, RFIDs, to more sophisticated mobile user devices such as smart phones, PDAs, and even notebooks, is the preferred approach. In this context, the network has a very strong ad hoc nature, only occasionally requiring communication to the existing network infrastructures (Internet backbone and cellular networks) whenever the need arises.

The architectural requirements are somewhat similar to those addressed in the European Union's BIONETS initiative [1], which considers service-triggered networking, in which "communities of interest" of users determine the services to be run, and these services determine the underlying communication information format and protocols necessary to support them. In the home network domain, however, the scaling requirements are not as central and the user communities much more static than in the general framework adopted in the BIONETS project. For (notational) convenience, we will adopt the same embedded autonomic peer-to-peer model based on localized and occasional information exchange, supporting a large range of device capabilities, ranging from low-resource tiny devices (called

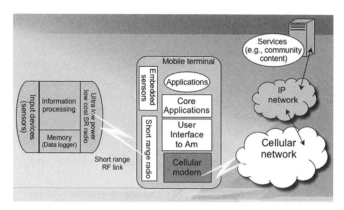

FIGURE 11.1   A mobile terminal as a gateway to both local and global information.

T-Nodes, e.g., sensors, actuators, RFIDs) to more sophisticated mobile user devices (called U-Nodes, e.g., PDAs, notebooks, phones, etc.).

For illustrative purposes, we will consider an architecture where a mobile terminal (i.e., a U-Node) plays a central role and acts both as a gateway to the local (more static) sensor networks (typically consisting of mostly T-Nodes) and as a high-power sensor node that is also capable of collecting sensor information. Choosing a mobile phone for such a central role is quite natural—it is the trusted device that a user always carries with him or her, and thus it can act as gateway collecting and distributing information not only from static networks but also from other mobile phones (Fig. 11.1).

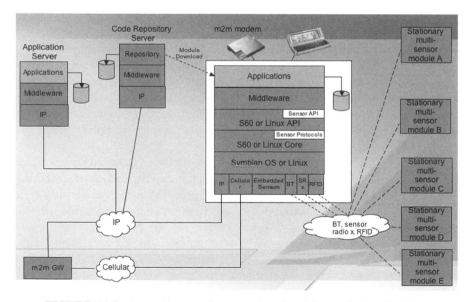

FIGURE 11.2   An architecture for sensor-based mobile terminal applications.

One possible more detailed architecture is illustrated in Figure 11.2, where the mobile terminal has capabilities to communicate with various alternative radio technologies such as Bluetooth [2], RFID [3], ZigBee [4], Wi-Fi [5]. In this modular architecture, the application code can be downloaded dynamically to any platform running either Symbian, Windows Mobile, Palm, or Linux platform, all of which have special sensor APIs for various stationary sensor types.

In order to illustrate the possibilities of such an approach, in the next section we will discuss one novel application category that can be built on this heterogeneous architecture. We will call this "spotting."

11.2 "SPOTTING"

In homogenous sensor networks, many of the applications rely mostly on one-way communication to the sink or gateway node. This is quite natural when one considers for example biomonitoring of a forest or detection of moving objects and the very limited resources of the (T-type) sensor nodes. In the home environment, applications tend to be interactive and require information flow in both directions. This is particularly true when a mobile terminal is viewed as a sensor node, and to be able to justify the use of mobile terminal resources, the user expects to see direct benefit typically in the form of a service provided by servers in the local or IP network. This interactive, service-oriented approach characterizes much of the sensor network computing for the home environment, from remote monitoring of the elderly to intelligent monitoring and possibly control of appliances and devices to management of entertainment media.

11.2.1 Tagging Physical Objects: "Spots"

Have you ever wanted to leave a reminder note for other family members but do not have Post-it or to pen available? or to leave a personal note that you do not like other than some family members to read? to leave an active "watcher" on your lawn that reminds when it needs trimming?

Tagging locations is one of the most interesting concepts in location-based services, which have become available due to the development of mobile positioning techniques. Companies like TagandScan offer users "virtual grids" where data can be attached. The accuracy of such grids varies from 100 meters to several kilometers due to the inaccuracies of the positioning techniques. A "spot" is a generalization of this "location-tag concept." As opposed to being attached to a location, a spot is a virtual object that is attached to a physical object (which can in a special case be a fixed part of a location). Here our interest lies in objects in the home environment, but the concept is of course not restricted to home domains. The virtual object can be a text message, sometimes with a special meaning such as a Web address or a phone number, an image, or a small (Java) program. The presence of spots can be "sensed," and they can be read or executed by a mobile terminal equipped with "spotting software." In some cases, the object can be seen superimposed on the streamed camera image; sometimes it has a symbolic form such as an icon that appears if a spot is sensed. In simple terms, a tagged object (also known as a spot) is made available to the intended party when he or she is in the vicinity of a wireless sensor node (which can be a tiny sensor, a Wi-Fi access point, or some other type of wireless access node) to which it has been physically attached. It should be noted that a spot can also be virtually attached to one or more wireless access nodes (including sensors) at a global level to one or more intended recipients by taking advantage of the underlying IP network. The advantage of spotting is also that it is nonintrusive and thus does not "spoil the visual scene." Digital Graffiti is naturally a special case of spotting. Although one can envisage a wide range of use cases, including location-aware advertising and messaging spots, in the remainder of this chapter we focus on an image object as an example of a spot to explain how such an application might work.

This image application creates a catalogue of objects anywhere, for example, a home, for later retrieval to find out where it is placed and its associated features. This application could be used to search for a lost or a forgotten object if it has not been moved from its originally stored location. To save the spot, the user takes a picture of the object from a handheld device and stores the picture and/or its extracted features computed from a Key function (described in the next section), any other attributes of interest, its location, the user ID, permissions who can access it, and even the geographic area(s) from where this object could be stored, all in a database. This set of parameters (called a tuple) can be later retrieved by a query, which can be direct or indirect (described in the next section). In a similar example, where RFID is used to tag an object, the tagged parameters as illustrated above are stored in a database. When the user wishes to retrieve the same, he or she simply scans the area of interest with an RFID reader or sensor, and when the object is in its vicinity, all its parameters (along with its picture) are retrieved and displayed to the user.

11.2.2 Spot Operations

In order to simplify the presentation, we need to introduce some basic notation to shorten the descriptions:

- Key image: K_img
- Key function (surjection): K: K_img -> K_id
- Key data: D
- Key location: L
- User ID: U
- Tuple (K_id, L, U, D): *Spot*.

The basic operations to be supported are naturally "Spot saving" and "Spot retrieval." The latter operation, however, can either be "Direct retrieval" or "Indirect retrieval" where the indirect retrieval involves additional search operation to just plain key identification.

11.2.2.1 Spot Saving

1. User focuses his or her camera to a region of interest to lock the key image.
2. User specifies the data to be attached to the object.
3. Terminal computes the K_id from the key image using the key function.
4. Terminal obtains the location information.
5. Terminal forms the spot tuple from user ID, key ID, data and location.
6. Terminal sends the tuple to the spot server, which is either local to the location or lies somewhere in the IP network.

In case of RFID sensing, step 1 is replaced by reading the RFID tag attached to the object and step 3 becomes trivial.

11.2.2.2 Spot Retrieval

As pointed out earlier, there are several different use cases for retrieval, depending on the key used for querying. We distinguish between exact or *direct retrieval* and *indirect retrieval* or search. For direct retrieval, we may use either K_id, L, or U as the query. For indirect retrieval, we may use any piece of information (or an aggregate) as the query, viz. K_id, L, U, D.

Obviously, the direct retrieval may be implemented as a simple database lookup. Indirect retrieval is in effect a content-based search operation, in which ranking plays a crucial role. One possible candidate is discriminative ranking called AinoRank, which roughly works as follows.

Given a query, retrieve all tuples satisfying the query. Let's denote the resulting set with Q. Score all the attributes according to how well it discriminates Q from the rest of the tuples. Score each tuple by summing scores of its attributes. Order the tuples by descending score. The spot with the top-most score or the spots with the top n scores can now be retrieved and displayed. For more details of this content-based ranking approach, see, for example, Refs. 6 and 7.

Direct Retrieval

1. User focuses her camera to a physical object to lock the key image.
2. User gets the attached data from the server.
3. User requests all the attached data, given her current location.
4. User requests all her own information (or in the general case information from the community members, for example, information left by people in the "buddy list").

Indirect Retrieval

1. Location-based search: User requests all the attached data, given his or her current location (as with direct retrieval [step 2]). The server ranks all the other locations with respect to this location (set of spots).

    Example motivation for such a query: "Give me all spots (especially from other locations) resembling my current location L."

2. Data-based search: Given some attribute(s) of data (a word, an image feature), rank the data in accordance with the attribute(s).

   Example motivation for such a query: "Give me all spots with this kind of information."

3. User-based search: Given a user ID U, rank all the data in accordance with the user's tuples.

   Example motivation for such a query: "Give me all spots (esp. other users, locations) resembling this user (myself)."

4. Key-based search: Since key_id $->$ data mapping is not unique (but $(K\_id, L, U) -> D$ is), we may have multiple users and locations corresponding with the same K_id, thus a K_id spans a set of tuples. This allows us to rank all data in accordance with K_id.

   Example motivation for such a query: "Give me all spots which are attached to an image resembling this one" or "Give me all the users who have used a spot like this."

11.2.3   On Key Function K

It is quite evident that the key function K plays a central role in spotting—a bad choice of key function might result in many mismatches of images about the same object as well as confusion of different objects in the retrieval. For good key functions, a user can focus his or her camera on a region of interest, take a picture, and attach some data to it. The next day, the user repeats the act and retrieves the same data. This should work even though the camera orientation varies wildly, illumination conditions vary, and even the camera may not be the same (e.g., the same user uses a different mobile terminal). A good key function should make this possible.

It is not possible here to go into the details of the various properties "spotting" imposes on the viable key functions. Therefore, we only list some basic requirements that need to be satisfied. Some of them are implied by the identification problem and some by the architecture in Section 11.1:

1. Illumination invariance.
2. Rotation invariance (not as strict as the previous).
3. Resolution invariance (not as strict as the previous), to allow changing cameras.
4. Real-time computing (i.e., computationally cheap).
5. Preferably yields a confidence measure, so that the user can be warned for an "unstable" spot.
6. Resulting K_id short (i.e., at most about a kilobyte so not to create a too-large bandwidth overhead).

One should observe that for home domains, some of these properties are easier to achieve than in the general case. Illumination variation is typically less severe, physical objects tend not to be changed as much (e.g., by painting), and the changes tend to be slower over time. Similarly, as only a restricted group of individuals has access to the physical objects at home, the indirect retrieval search problems become easier compared with spotting in public places and/or on public objects.

Interestingly enough, construction of such a key function can utilize techniques from information theory, such as lossy compression, hash functions, fingerprinting or Tversky similarity, sparse coding, and so forth (see, e.g., [8]).

11.2.4 Spotting with Additional Sensor Information

Up to this point, we have discussed spotting that is based on the mobile terminal also functioning as a sensor (i.e., camera, location information) and a gateway. This basic scenario can be enhanced in several ways by adding static sensors in the heterogeneous architecture.

First of all, some of the static information can be used for a much finer grained location L identification by using fingerprinting. For example, various short-range sensors that are only visible to the mobile terminal when they are in range are particularly good for such purposes. Notice that in general signals from (fixed) Bluetooth devices are not very useful in this respect because in normal home environment, their signal can be picked up anywhere.

Second, adding sensor information as part of the data in a spot opens interesting possibilities for various types of diagnostics and is particularly useful for transient failures (a "service spot") where both visual and sensor information such as temperature can be stored to be retrieved by the service personnel coming to fix the problem. With the popularity of all types of social networks, we are entering an era of participatory Internet and content; consequently, the idea of spotting, which already embeds in it the location information, among others, can unleash a range of new services and applications where spots can be shared based on the locations of the members of the social group resulting in much richer content.

11.3 CONCLUSIONS

In this chapter, we have discussed heterogeneous wireless sensor networks at home and the advantages of such deployments for pervasive computing. The architectural viewpoint taken was based on viewing the mobile terminal acting both as a gateway to local static sensor networks and as a dynamic sensor node. We then presented in more detail one of the consumer-centric new application categories called "spotting" (i.e., adding virtual objects to the physical world) and how it would utilize the mobile terminal centric viewpoint to sensor networks.

ACKNOWLEDGMENT

The work presented has benefited from discussion of various colleagues at NRC, in particular Ville Tuulos, Jukka Perkiö, Jan Bosch, and Petteri Saarinen.

REFERENCES

1. Chlamtac, I., Carreras, I., Woesner H., From Internets to BIONETS: Biological Kinetic Services Oriented Networks. The Case Study of Bionetic Sensor Networks, In *Advances in Pervasive Computing and Networking*, B. K. Szymanski and B. Yener, Eds., Springer Science+Business Media, New York, 2005.

2. Available at www.bluetooth.com.

3. Available at www.rfidjournal.com/.

4. Available at www.zigbee.org.

5. Available at www.wifialliance.com/.

6. Perkiö, J., Tuulos, V., Buntine, W., Tirri, H., Multi-Faceted Information Retrieval System for Large Scale Email Archives. In *Proceedings of the IEEE/WIC/ACM Conference on Web Intelligence (WI 2005)*, pp. 557–564, 19–22 Sept., 2005, Compiègne, France.

7. Tuulos, V., Silander, T., Language Pragmatics, Contexts and a Search Engine. In *Proceedings of the International and Interdisciplinary Conference on Adaptive Knowledge Representation and Reasoning.*, Espoo, Finland, June 2005.

8. MacKay, D., *Information Theory, Inference and Learning Algorithms*, Cambridge University Press, Cambridge, 2003.

INDEX

A interfaces, 95
A/V coding, 52
A/V format variants, 67
A/V object, 65
AAA. *See* authentication, authorization and accounting server
AC-3 audio format, 32
Academic research, related work and, 74
Access Control List (ACL), 129
ACL. *See* Access Control List
Adaptive content verification, 141
Adobe, 143
Advanced Micro Devices (AMD), 18
AI. *See* air interace
Air Interace (AI), 116
Air Station OneTouch Secure System (AOSS), 123
AirG, 8–9
AirStation OneTouch Secure System (AOSS), 130
Algorithms, cryptographic, 139
AMD. *See* Advanced Micro Devices
AOSS. *See* Air Station OneTouch Secure System
Apple Computers, 16, 164
  Quick Time, 58

Application convergence, 12–14
ARIB, 51
ATSC, 51
Audio coding, 52
Audio coding formats, 30–32
  AC3, 32
  Dolby Digital, 32
  Linear Pulse Code Modulation (LPCM) format, 31–32
  mobile music services, 32
  mobile music services, XM Radio, 32
  MP3, 30, 31
  MPEG-2 ACC, 31
  MPEG-4 HE-AAC, 31
  RealAudio, 30, 31
  summary of, 31
  Windows Media Audio (WMA), 30, 31
Audio format variants, 56
Authentication, 137, 140
  Authorization and Accounting (AAA) server, 96

Bebo, 8
BL. *See* Broadcast Flag
Blogging, 8

Technologies for Home Networking. Edited by Sudhir Dixit and Ramjee Prasad